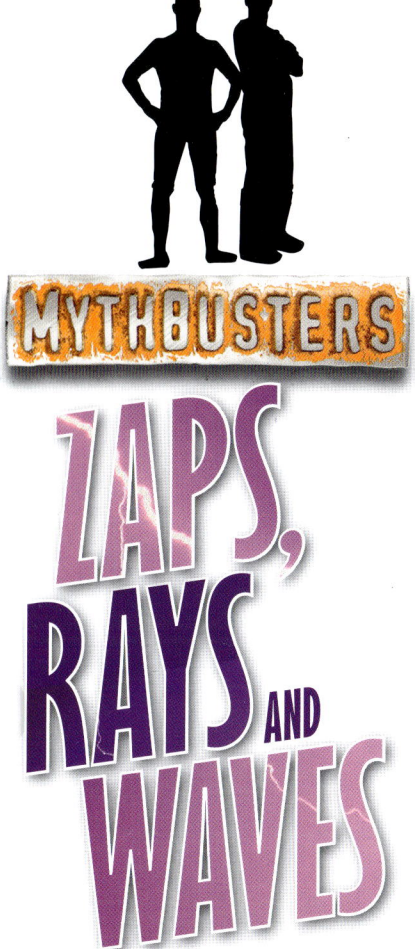

MYTHBUSTERS

ZAPS, RAYS AND WAVES

TECHNOLOGY, ELECTRICITY AND THINGS OF A WAVELIKE NATURE

NICHOLAS SEARLE

Published by Wilkinson Publishing Pty Ltd
ACN 006 042 173
2 Collins St
Melbourne Vic 3000
Ph: (03) 9654 5446

www.wilkinsonpublishing.com.au

Copyright © Beyond Entertainment Ltd 2007

First published 2007

All rights reserved. No part of this publication may be reproduced, stored in a retrieval system or transmitted in any form by any means without the prior permission of the copyright owner. Enquiries should be made to the publisher.

Readers should not engage in the activities described in this book. Readers performing the activities in this book do so at their own risk and discretion. Every effort has been made to ensure this book is free from error or omission. However, Beyond Entertainment, Wilkinson Publishing, the Author, the Editor or their respective employees or agents, do not warrant or endorse the activities described in this book and specifically disclaim any responsibility for injury, loss, or damage occasioned to any person acting on or refraining from action as a result of the material in this book whether or not such injury, loss or damage is in any way due to any negligent act or omission, breach of duty or default on the part of Beyond Entertainment, Wilkinson Publishing, the Author, the Editor or their respective employees or agents.

National Libraries of Australia
Cataloguing-in-Publication data:

Searle, Nicholas
    Mythbusters zaps, rays and waves : technology, electricity
      and things of a wavelike nature
    ISBN 9781 9213 32074 (pbk.).

    1. Mythbusters (Television program). 2. Science - Experiments.
    3. Belief and doubt. 4. Urban folklore. I. Title.

    507

Page and cover design by Spike Creative Pty Ltd.
Prahran, Victoria. Ph: (03) 9525 0900.
www.spikecreative.com.au

Printed in China.

About the Author

In his mind he stands nearly 200cm, is a muscular 100kg, speaks four languages and plays the trumpet like Miles. In truth, Nicholas Searle is shorter, heavier, more linguistically stunted, and can barely fumble through *It Must Be Love* on a bass guitar.

Nicholas has worked on a number of TV science programs, including the Discovery Channel's *Mythbusters* and *One Step Beyond*, plus the Australian Broadcasting Corporation's *Catalyst* and *The New Inventors*. He's also written children's animation, documentaries and stand-up comedy.

Literally surrounded by PhDs (in Biology, Psychology, Architecture, Social Geography, Post-Modern Literature and Cultural Theory), Nicholas has so far managed not to acquire one, but he has the utmost respect for scientific knowledge and its potential for quality laughs.

Warning! Read This!

DO NOT attempt these experiments, tests, trials, or any activity in this book at home, work, or anywhere else for that matter. The Mythbusters Team is trained, experienced and has resources at its disposal which include safety and medical teams.

Contents

Introduction — **08**

The Mythbusters ... who are they? — **12**
You've been seeing them in your living rooms for years ... now find out who they are, and what makes them tick.

1. The Basics of Zap — **24**
We might take electricity for granted, but the Mythbusters boys try to shed some light on what is an amazing resource.

2. The Basics of Rays — **66**
Some call them rays, others call them waves ... and they're kind of both right. No, we're not making this any easier, are we?

3. The Basics of Waves — **100**
Waves are fun, right, as long as you have a surfboard ... but we reckon the other kind can be pretty entertaining too.

Introduction

We'd like to welcome you to what promises to be an hilarious but cool introduction; an introduction that we hope might one day be regarded as one of the greatest of the 21st Century.

The introduction is here to explain (hilariously, yet cooly) the reason for the existence of the book it introduces (and the others that will follow).

This book is in your hands, first, because you've just purchased it (or are soon to do so) but, more importantly, because of the perspicacity, probity and above all popularity of a TV science program that is even <u>more</u> likely to be regarded as the greatest of the 21st Century than this introduction.

Any guesses as to what that program might be?

The answer is, of course, *Mythbusters*; a colossus among giants; the first, the best, the greatest television program ever to bring the steely experimental probes of truth to the stagnant, sordid darkness of urban mythology.

So let us begin by congratulating you on the wisdom and value of the purchase you've just made. Can we tell you that it suits you? It certainly does – it even matches your eyes.

Tell you what; why don't you take it out of the bag and let other people see what you've just bought? A new purchase of any kind gives you a buzz, and much more so when the purchased item makes you stand out from a crowded bus, mall or university lecture theatre like a lemming who decided not to jump [0].

In fact, why don't you stand up right now and say the following in a loud, proud and steady voice: "I've just purchased a copy of the *Mythbusters* book *Zaps, Rays and Waves; Technology, Electricity and Things of a Wavelike Nature*, and I feel fan-<u>tastic</u>!"

Are they looking?

Of course they are! And so they should, for this book sets you apart from the mob. You are no longer willing to live with assumption and guesswork – nothing but incisive science and regimented logic pierced by hilarious gags and witty badinage will satisfy you.

Be proud and be brave, stolid purchaser, for the world will never be quite the same again. Let's bust some myths!

Actually, not just yet … **Caveat Lector!**

In creating this book, strenuous efforts have been made to verify both the reliability of the facts and the humorosity of the jokes.

To the best of our understanding the facts are true and the jokes are funny.

HOWEVER.

Just as this book should not be your sole reference for a fourth year university high-density physics open book examination, neither should it prop up your entire conversation during a cocktail or dinner party.

If you would like to learn more about science, then get out there and BUY MORE BOOKS ABOUT SCIENCE. Get on the web and seek out a version of the truth that suits your worldview. If you would like to learn how to use humour to appear more interesting, then get out there and HANG WITH AMUSING PEOPLE WHO'VE HAD FUNNY LIFE EXPERIENCES.

[Editor's Hint – any material with *Mythbusters* in its title will cover both.]

Ab Initio!

[0] Yes, we start from '0' here (if it's good enough for the Laws of Thermodynamics, it's good enough for us). With that, welcome to the footnotes! This is where the book reaches into the misty crevasses of the mind to retrieve footnotable thoughts such as the following: "Yes, we know that lemmings don't really jump off cliffs, and yes, we know that (and are rather amused by the fact that) this 'mythbusting' book has started by repeating what is demonstrably a myth. Why not look out for others as we progress? Don't be afraid to stand up and shout them out, no matter where you're reading!" How was that? Enjoy it? Good – plenty more where that came from.

The Mythbusters...
who are they?

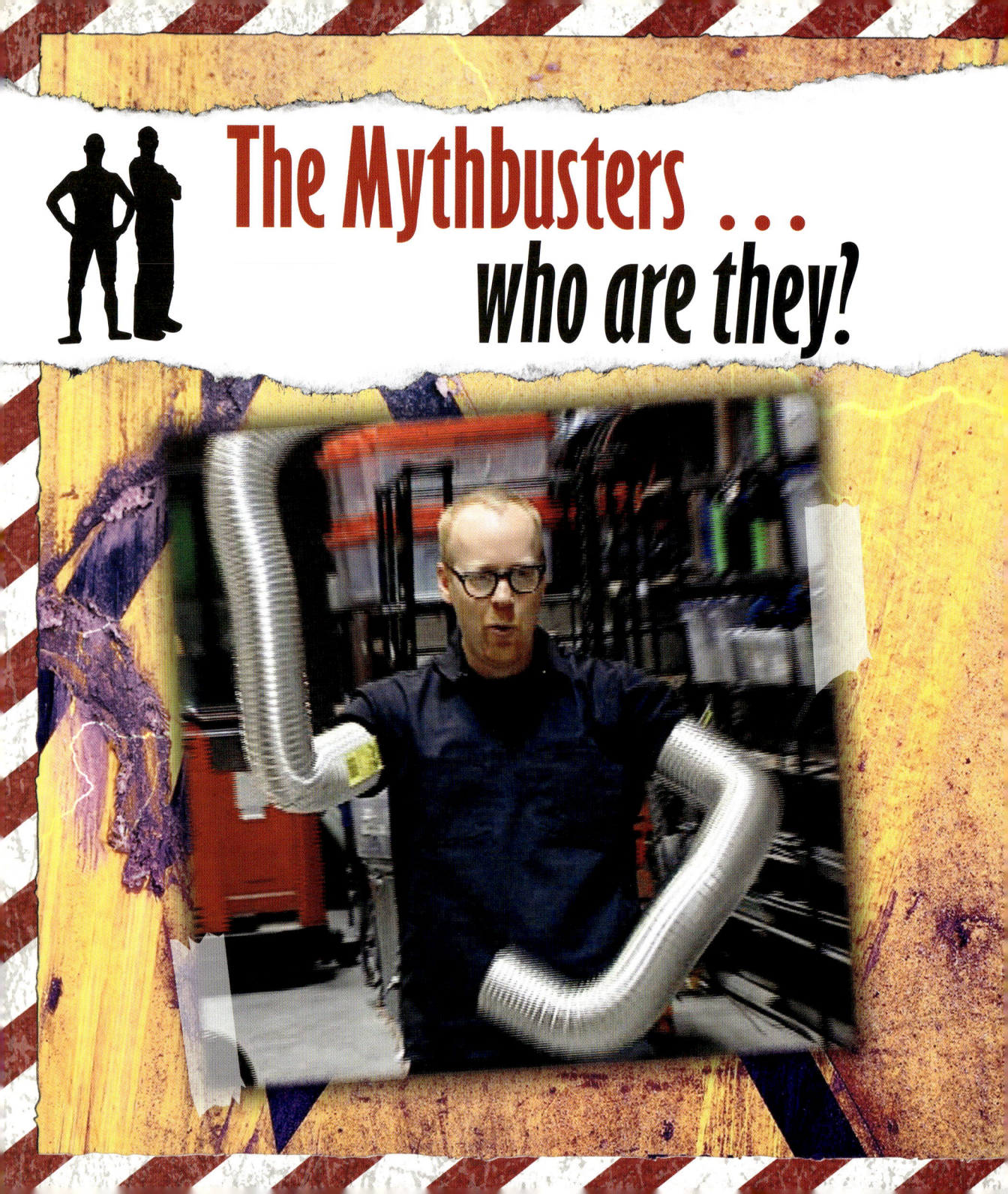

ADAM SAVAGE

You know Adam, he's the one with the glasses. Okay, so both he and Jamie wear glasses, but Jamie's look studious and kind of inoffensive, Adam's are, well either uber-geeky or uber-cool. We just can't decide.

For some reason, it's Adam who is regularly putting himself in harm's way. Again, it makes for an interesting comparison with his co-host. Jamie tends to think things through … he admits to spending plenty of time when the cameras are switched off, thinking through a solution for the Mythbusting problem.

And he's usually the one who thinks about the thing called 'consequence'.

Adam has adopted a 'let's just do it' mentality. Which means he lives on the edge a bit more … which is kinda fun.

See, Adam always made stuff. That's his thing. But instead of paintings and cardboard boxes, Adam was constructing spaceships and Buddhas, puppets to rifles, sculptures and working toys ... and just about anything else he could think of.

He has taken that philosophy into his adult life, creating a CV which might not be traditional, but sure is interesting.

Let's count through the various jobs … animator, graphic designer, stage and interior designer, carpenter, welder and scenic painter. He's worked in everything from metal to glass, plastics to injection molding, and pneumatics to animatronics.

And for the past eight years, he's concentrated on the special-effects industry. His handiwork has been seen on more than 100 TV commercials, and a dozen or so feature films, including *Star Wars Episodes I and II*, *Space Cowboys*, *Galaxy Quest*, *Terminator 3* and *The Matrix* sequels. He has done research and development for toy companies, acted in commercials and films, and done props and sets for Coca-Cola, Dow Corning, Hershey's, Lexus, and a host of New York and San Francisco theatre companies. Adam is also a sculptor, of mixed media assemblage, whose work has been represented in more than 40 shows in San Francisco and New York.

JAMIE HYNEMAN

Jamie might look like the thoughtful one – okay, that's him alright – but there's a dark side to the *Mythbusters* co-presenter. A visual effects expert and founder of M5 Industries, a special effects workshop where *Mythbusters* is often filmed, he is also known in the *BattleBots* circles for his robot entry, Blendo.

So, maybe the Indiana farm boy likes to let a little steam off every now and then.

Like most of the *Mythbusters* team, there are plenty of strings to his bow: wilderness survival expert, boat captain, diver, linguist, animal wrangler, machinist and chef, to name a few. His career has been equally diverse: Jamie earned a degree in Russian languages and literature before he moved over to the visual-effects industry.

Once he had joined that field and had worked for several production companies, Jamie found his way to Colossal Pictures' model shop, where he managed the production of models and special effects for hundreds of commercials and movies. Then, eight years ago, Jamie took over the shop and created M5.

A distinctive part of Jamie's appearance is his vicious moustache, dark beret and white lab shirt, and he is known for his calm, logical, no-nonsense demeanor. Which makes him a nice counter to Adam, if you know what we mean.

Jamie has also worked on commercials for major automobile manufacturers, soft-drink companies (including 7-UP) and athletic shoe companies (including Nike). And in the midst of all this activity, he has diversified his company into toy prototyping as well.

THE MYTHBUSTERS ... WHO ARE THEY?

ZAPS, RAYS AND WAVES

KARI BYRON

She didn't know it at the time, but Kari has been practising to be a Mythbuster since she was, oh, about five.

By the age of five she was setting up experiments to test on her sister and using dolls as crash test dummies. Luckily for her parents, they always caught her right before little sister took a ride down a laundry chute or was the subject of an "around-the-world" attempt on the playground swings.

Kari studied film and sculpture at San Francisco State University where she wrote, directed and starred in B-level schlock-horror films, sculpted intricate model dragons for *Dungeons & Dragons* fans, and graduated as an artist, working in sculpture and painting. Kari has had successful exhibitions at some of San Francisco's leading galleries.

Kari also has worked in a kosher bagel store, and has posed as a store mannequin to foil shoplifters … not glamorous, perhaps, but let's just say it required good discipline.

"Artist" was only one of many hats she wore while searching for her place in the world. Her sculpting skills and love for odd jobs soon led her into the field of model-making and toy-prototyping, which led to a job with Jamie at M5 Industries. It was at M5 that Kari got her first big break with the *Mythbusters* team.

During the "vacuum toilet" segment of one of the first episodes (which examined whether a person could get sucked into an airplane toilet), Jamie needed a 3D scan of a person's backside, and Kari had the right … well, you get the rest.

Basically, she was in the right place at the right time. The rest is history. Who knew that a degree in film and sculpture would actually be applicable to a real-life career one day? Now Kari works with the *Mythbusters* team, using science and Yankee ingenuity to solve the mysteries of today's most compelling urban legends.

GRANT IMAHARA

You can usually recognise Grant by the incredulous look on his face, having just been asked by Jamie or Adam to construct something, well, preposterous.

See, they look to Grant for the electronics expertise … when you want to make something go bang, it's not a good idea to get too close, so one of Grant's jobs has been to make things go bang from a safe distance.

Which is fine… he's sort of qualified for it, being a radio-controlled expert (well, an expert in radio control things, then).

Grant was a former animatronics engineer and model-maker for George Lucas' Industrial Light & Magic, where he worked on the movies *The Lost World: Jurassic Park, Star Wars – The Phantom Menace, Terminator 3, A.I. – Artificial Intelligence* and *The Matrix Reloaded.*

And, yes, before you ask the question, he was one of the few people licensed to operate R2-D2. No, seriously. Grant also developed a custom circuit to cycle the Energiser Bunny's arm beats and ears at a constant rate, and is responsible for all the electronics installation and radio programming on the current generation of bunnies. And he continues to serve as the bunny's driver and crew supervisor on numerous commercials.

He shares Jamie's love of robots, too, having written the book, *Kickin' Bot: An Illustrated Guide to Building Combat Robots.* His own machine, Deadblow, is a *BattleBots* champion that strikes terror into the hearts of fellow competitors.

TORY BELLECI

Well, if there's been a Mythbusting intern by definition, we think it was probably Tory. He kinda walked in off the street in 1994, wandering into Jamie's workshop looking to tinker, learn … and maybe get some work if it was going.

Surprise, surprise … more than a decade later, he's an integral part of the team.

When he doorknocked Jamie as a young film student, Tory was looking to start a career in special effects. Naturally, Jamie put him to work sweeping the shop floor.

But Tory quickly moved up the ranks. Three years later, he landed a job with George Lucas' special effects division, Industrial Light & Magic, where one of his biggest projects was building models for the *Star Wars* trilogy. The Federation Battleships and Podracers in *The Phantom Menace* and *Attack of the Clones* are examples of his work. Um, wow!

Tory's other special-effects work can be seen in the *Matrix* trilogy, *Van Helsing*, *Peter Pan*, *Starship Troopers*, *Galaxy Quest* and *Bicentennial Man*. One of his short films has appeared in the Slamdance Film Festival and on the Sci-Fi Channel.

When he's not busting myths, Tory's busy dreaming up his own ideas for films and TV shows. One of his short films has already appeared in the Slamdance film festival and on the Sci-Fi Channel.

Tory was born and bred in Monterey, California, right near the famous Pebble Beach golf course. He once shot a blistering 15 over par round of 87 there … then decided special effects was more his game.

"Jamie is known for his calm, logical, no-nonsense demeanor. Which makes him a nice counter to Adam, if you know what we mean."

THE MYTHBUSTERS ... WHO ARE THEY?

ZAPS, RAYS AND WAVES

23

The Basics of Zap

Chapter 1

Electricity is now such an everyday thing that you can easily stroll through life and barely think about it – that is, until it zaps a million volts through your tongue stud.

Electricity is, in fact, excitingly dangerous stuff; amazingly useful as well as utterly fundamental to the basic nature of the Universe in which you're currently reading this book. Oh yes, electricity is big-time important.

What do we mean by fundamental? To explain, just take a moment to look at your little finger, which is made up of various elements from arsenic to zinc [1].

Of all these elements let's choose the simplest: Hydrogen.

Inside one single hydrogen atom in your little finger is a nucleus surrounded by an electron. *A what surrounded by whats?*

Okay … an atom is a very, very small thing that clumps together with others of its kind to make up lots of other things … such as everything in the Universe.

Inside the atom of hydrogen is a nucleus; a blob of stuff otherwise known as protons and neutrons (one of each in the case of hydrogen). The proton and neutron just kind of sit there [2], but swinging around the nucleus they create by 'just sitting there' is a smaller blob of stuff that spins around the big blob of stuff.

"Come on! How hard can it be to get a light bulb to light?" – Adam

This smaller blob is an electron, and the electron spins around the nucleus for a very good reason … they're in love.

Electrons enjoy being particles, but at the same time they get a kick out of being waves. Eh? This mindbender is given the term 'wave-particle duality' by scientists who hope one day to understand why it should be. The impractical upshot is that everything around you – and you yourself – is really made up of waves of energy. So don't panic; next time you drop a glass and it smashes to pieces, take comfort in the fact that it was probably all those waves that made your hands slippery, and the slivers of pointy glass on the floor are just a bunch of waves themselves, and won't hurt anybody.

Okay, that's a little misleading; the term scientists prefer to use is 'attraction', but you know what they're really talking about (scientists can be a little shy about the 'L' word).

The reason the electron is attracted to the nucleus is because they have opposite charges, positive in the case of the nucleus (because of the proton [3]) and negative in the case of the electron.

And thus is understood the well-known saying 'opposites attract'.

Sure, atoms are small, but the little bits of stuff (oh, alright, 'matter') inside them are even smaller. If an atom was the size of a football stadium, then the nucleus would be a marble in the middle. An electron, however, would be smaller than a speck of dirt – IF IT HAD ANY DIMENSION AT ALL which is – apparently – possible. That leaves our 'football stadium' atom pretty much empty, and when you add the fact that there's also bags of space <u>between</u> atoms, you could be forgiven for thinking that this book, your hand, the chair that's holding you up off the planet, and even <u>the planet itself</u>, are mostly empty space. But don't go testing it out by driving your car at a tree or anything – there's still enough matter to make a fair dent in your insurance policy.

[1] Yes, we're serious – for more freaky body factoids check out the next title in the series, 'Rock the Body'.

[2] In fact, the proton and the neutron are made up of quarks, which do dance around a bit, as would you if you were named after a cheese. For more information, get yourself a Difficult Degree in Complex Worthwhile-iosity from your local College of Advanced Cleverness.

[3] It's the proton that carries the positive charge. Neutrons have a neutral charge, as the name suggests. Isn't it nice when scientists do that?

And so all electricity is a result of the spunky electric charge of tiny, tiny, tiny, tiny things (or 'particles' for those seeking to impress).

"All right!" – I hear you say – "Now that I know all about electricity I'm going to wire up my guitar amp to a lightning rod and shock my mates with a killer power chord!"

Good luck with that, although it might be you who gets the shock. Not surprisingly, there is more to electricity than a few tiny, circulating blobs, no matter how fundamental they are.

For example, the best parts of chemistry are based on the fact that under the right circumstances (flattering lighting, French champagne, New Zealand 'Bluff' oysters) atoms from different elements sidle up to one another and – ahem – 'share electrons'.

This kind of activity leads to the formation of really useful things that you probably use every day despite the warning of your cardiologist, like table salt (sodium chloride to the rest of us) in which the sodium atoms have shared an electron with the chlorine atoms, and their strong electrical attraction forms a new solid material that tastes great in everything from guacamole to a lightly grilled salmon fillet.

Admittedly, we're talking about tiny, tiny amounts of electricity here, not even enough to make your brain summon enough zap in your nerves to pull the muscles in your fingers that will turn the next page.

However, the principle of electrical attraction between electrons, protons and neutrons is vital for the stability of the universe – and we're all in favour of that, aren't we?

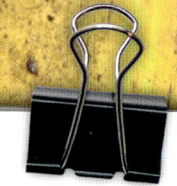

The term that describes these phenomena, and much, much more besides, is 'electromagnetism'. So, major points to those readers who guessed that all this 'attraction' talk, along with stuff about 'opposites', 'positive and negative' and 'guacamole' meant that magnetism had to be along for the ride (how else would you attach that spunky Mexican girl's recipe for guacamole – and her phone number – to the fridge?).

Thankfully for the viability of computer games, electric guitars and the internal combustion engine, electricity can do more than just plug together the right elements to spice up your French fries (or that guacamole).

But the thing about electrons is that they can't be trusted to stick around. Some electrons take their relationship with a nucleus less than seriously, and it just takes a flutter of eyelashes and a saucy suggestion to make them park their negative charge somewhere else.

Adam and Jamie find this out when they investigated the mythical danger of a cell phone at a gas station [4].

STATIC AT THE GAS STATION

In the myth 'Cell Phone Gas Station' Jamie and Adam investigate the possibility that using your cell phone while filling your tank could cause an explosion. But in their search for the culprit, they soon dropped the cell phone theory and turned their attention elsewhere …

Jamie: "I think it's highly unlikely that a cell phone is gonna do this … it seems like a dumb idea to me."

Adam: "Well, actually … it seems that women are largely responsible for the gas station explosions, specifically because they get in and out of their cars while filling up. So they'll build up a static charge on their body, and then when they touch either the gas tank lip or the filler nozzle then that will ignite the gas and cause an explosion."

Jamie: "That seems more likely."

[4] - Why cellphone and not mobile? For the sake of clarity and a possible spike in American sales, (and given that Mythbusters is itself an Australian TV show masquerading as an American TV show), we'll masquerade as a Stateside publication and use the term 'CellPhone' instead of mobile phone.

[5] - In fact, it's called static electricity for two very good reasons; back in the first half of the 19th century scientists thought electricity came in five different flavours - 'static', 'voltaic', 'thermoelectric', 'magnetoelectric' and 'animal'. Despite their silly hats and complicated facial hair, these olde worlde scientists were a teensy bit wrong; the electricity they were describing was actually all the same stuff. So to cut a long story short (too late) we continue to use the term 'static' because (A) a static charge is an unmoving accumulation of electric charge (unmoving is what 'static' means – go look it up), and (B) because we're suckers for a bit of nostalgia. But if we're such suckers for the good ol' days, then why aren't there people floating around wearing silly hats and growing very long moustaches? The obvious answer to that is 'ask Jamie'…

If you've <u>never</u> generated a zap reaching for a doorknob after walking across a carpeted floor in your favourite brogues, then it's quite possible that you're living a virtual existence inside a giant computer programmed by pan-dimensional aliens who took little interest in the intricate details of life in our Universe.

So let's go ahead and assume that's not the case, and that you've experienced static electricity. It's all good fun when it's between you, the carpet and your dog's nose, but how does it actually work?

Well, it's all down to the flighty tendencies of those frisky, freewheeling electrons. If you can get enough of those electrons to jump off one material onto another material, you can build up a tasty little electric charge. For no good reason, this charge is called static [5].

All it takes is contact between two materials for electrons to jump from one material to another, so as you can imagine static can be a big problem for things with contacting parts of different and sensitive materials (think cameras, tape and record players, Apollo rockets …).

But a bigger problem for you and me and other innocent bystanders at the gas station is that the human body is really good at holding a static charge. When you consider that the things we wear such as silk, rayon and wool (okay, you don't have to wear them all at the same time) are really, really good at building up a static charge, it's a wonder we're not <u>all</u> making the evening news.

What Jamie and Adam found in 'Cell Phone Gas Station' was that people who get back into their vehicles when they were filling up their cars at the pump could easily build up a static charge in their body.

All it takes to build up a charge is to sit down in your car and stand up again. If the first thing you touch is close to the highly flammable fuel chugging into your tank, there's a good chance that little spark or static electricity you generated could start a fire. The real problems begin when you then try to pull the gas nozzle out …

The solution? Step away and ask for help.

'Grounding' (or 'earthing') describes the action of releasing an electric charge into something much bigger and safer than a puddle of petrol. The planet Earth is useful for this (hence the term) and it's what these gas station-exploding drivers are not doing before they reach for the nozzle, all charged up thanks to the layers of silk, wool and rayon they thought would be fashionable that morning.

So why do we get the spark anyway? The little spark that you see milliseconds before you blow up your local gas station is the result of the static charge that's built up in our body. There's a whole bunch of electrons clinging to you that now want to find their 'opposite'; that is, a place with an opposite charge.

The spark itself is actually little bits of the air between your finger and the gas pump 'ionising' as the static charge flows across them (an 'ion' being an atom that has recently lost or gained an electron). [6]

[6] So an ion is like a widow, or a wife, who's just lost, or taken a husband. Except it's one word for both. This isn't making things easier at all, is it?

[7] For more gruesome details about the body and electrocution, buy a copy of **Rock the Body**. Sure, you could just flick through it in the bookshop, but why take the risk that certain pages are connected to the mains with micro-wires bought from ex-KGB sources? Publishers will do anything for a sale these days, and a few suspicious-but-untraceable electrocutions would be headline grabbers. So go on, reach slowly for your wallet as you walk to the checkout, ensuring you keep one foot on the ground at all times and UNDER NO CIRCUMSTANCES LET GO OF THE BOOK.

Materials will tend to either give up or attract electrons when put in contact with other materials. Sometimes the transferred charge will be negligible, but at other times it'll be very apparent. This is called the Triboelectric Effect (named after 'tribos', the Greek word for 'rubbing'). In the myth 'Cell Phone Gas Station' Adam based his 'panty static generator' on this principle, and you can easily do something similar at home. If you've got a woollen jumper on a plastic hanger, give the hanger a rub on the jumper and check out the electricity you've generated by holding the hanger next to a slim stream of water from a tap. You should see the water move towards the plastic hanger!

A little spark that jumps – or 'arcs' – across a gap that's only 0.5mm across [7] might be 1500 volts. 1500 volts sounds like a lot, doesn't it? Then gibber at the fact that a spark that arcs across two or three millimetres would be something like 6000 to 10,000 volts.

And yet, strangely, unless it's arcing through flammable gases and turning your car into a news story, you'll only feel a little sting. Why?

One word – current.

It's all very well to be zapped by a big, chunky 10,000 volts, but without a high current you'll easily live to watch another episode of your favourite TV science series. Current and voltage go together like salt in guacamole. A pinch adds bite – a few handfuls and you're in coronary country.

So what is this 'current'? Just as a river's current is the flow of water in the river, electrical current is the flow of electricity through whatever it happens to be in at the time (powerlines, that old extension cord with the chewed plug, your heart…).

And as that river current can be gentle and quite pleasant to raft down on a warm spring morning, or so deadly fast that even your life vest won't save you from being trapped under a rock, so can an electric current.

If the volume of water in the river would be comparable to the voltage, then the whole enchilada – the current and the volume – is the overall power of the river. In electrical terms, this is measured in watts. Current is measured in Amps, or Amperes for the pernickety.

Just sitting around reading this book your body is operating at about 80 watts. To climb the stairs to get to your apartment to read the book, you 'powered up' to about 200 watts. The world-class sprinter that you always dreamed you would one day become (fat chance now, Tubby) might have been able to generate 2000 watts at the peak of their performance. It's all a bit Matrix-y when you think about it …

So, to continue the metaphor, you can paddle a canoe on the mighty Amazon without a hitch, or you can be swept off your feet when a fire hydrant bursts. When it's life and limb in the firing line, the current matters.

Luckily for all the undertakers out there, it doesn't take many amps to do you damage. With a bagful of volts behind it, you won't forget a zap with just a few thousandths of an amp (or 'milliamps'). With enough volts and a 'direct current' zap of, like, **half an amp**, you won't remember anything at all – you'll be dead.

That static spark hitting your dog's nose might only have one ten millionth of an Amp. But it needn't take much to build a dangerous charge – and Adam and Jamie were soon sparking off one another when they vied to create the biggest spark, as they confront the myth of the Static Cannon.

Direct Current? Direct Current and Alternating Current are not a reference to the greatest rock band Australia ever produced (except perhaps tangentially), but rather are different kinds of electricity. Direct Current is the old fashioned kind of electricity that flows in only one direction through a conductor – the kind of electricity you get out of batteries and the ends of fingers (as they approach the family dog). For the skinny on Alternating Current, read on to later pages, as this call out is already reaching maximum text density.

[8] Surprisingly, there are more myth titles with the word 'phone' in them than 'cannon' or 'jet'. Sure, we like phones, but … they're a bit blah – you know? However, if there was a myth about Exploding Jet Bullet Cannon Phones, you better believe they'd have a crack at it .

STATIC CANNON

Mmmmm … cannon…

Mythbusters love myths with the word 'cannon' in the title; also myths that refer to rockets, guns, bullets, jets, or explosions [8].

But 'cannon' myths are right up there. Adam and Jamie have made cannons out of trees, and steam (eh?), but the Static Cannon held as much – if not more – promise than any of them. A killer cannon that fires off a deadly burst of static electricity? Prepare the workshop!

Jamie: *"This is a really cool story. Lots and lots of static electricity."*
Adam: *"Potentially lethal, yes."*

And as lethal means 'capable of causing death' they'll be taking things pretty seriously. You'll see why soon enough … first the myth. Once upon a time on a building site, a builder found the need to reuse a large piece of PVC pipe. However, the pipe had been painted, so he began to sandblast the pipe to remove the superfluous paint.

Unbeknownst to the builder, the action of blasting all that sand along the pipe built up a massive static charge. When next he grabbed for it, the PVC pipe 'static cannon' zapped him stone dead.

Trust us when we tell you that Jamie and Adam <u>really</u> want to believe that this is possible, and they put everything giving the myth the best chance.

First, they dry out the testing zone, using sheets of plastic and painter's desiccant to create a moisture-lock. But why? Tiny droplets of moisture in the air form little escape rafts for all those electrons that Jamie and Adam need to maroon on the PVC pipe to build up a charge. Although the PVC pipe (PVC? Polyvinyl Chloride, of course!) wouldn't let current flow through it, you <u>can</u> build up a charge upon its surface by encouraging electrons to hop on (just as you could on that plastic coat hanger).

And how do they do that? The good ol' Triboelectric Effect, which in this case comes disguised as a mild-mannered, industrial-sized sandblaster (and plenty of sand).

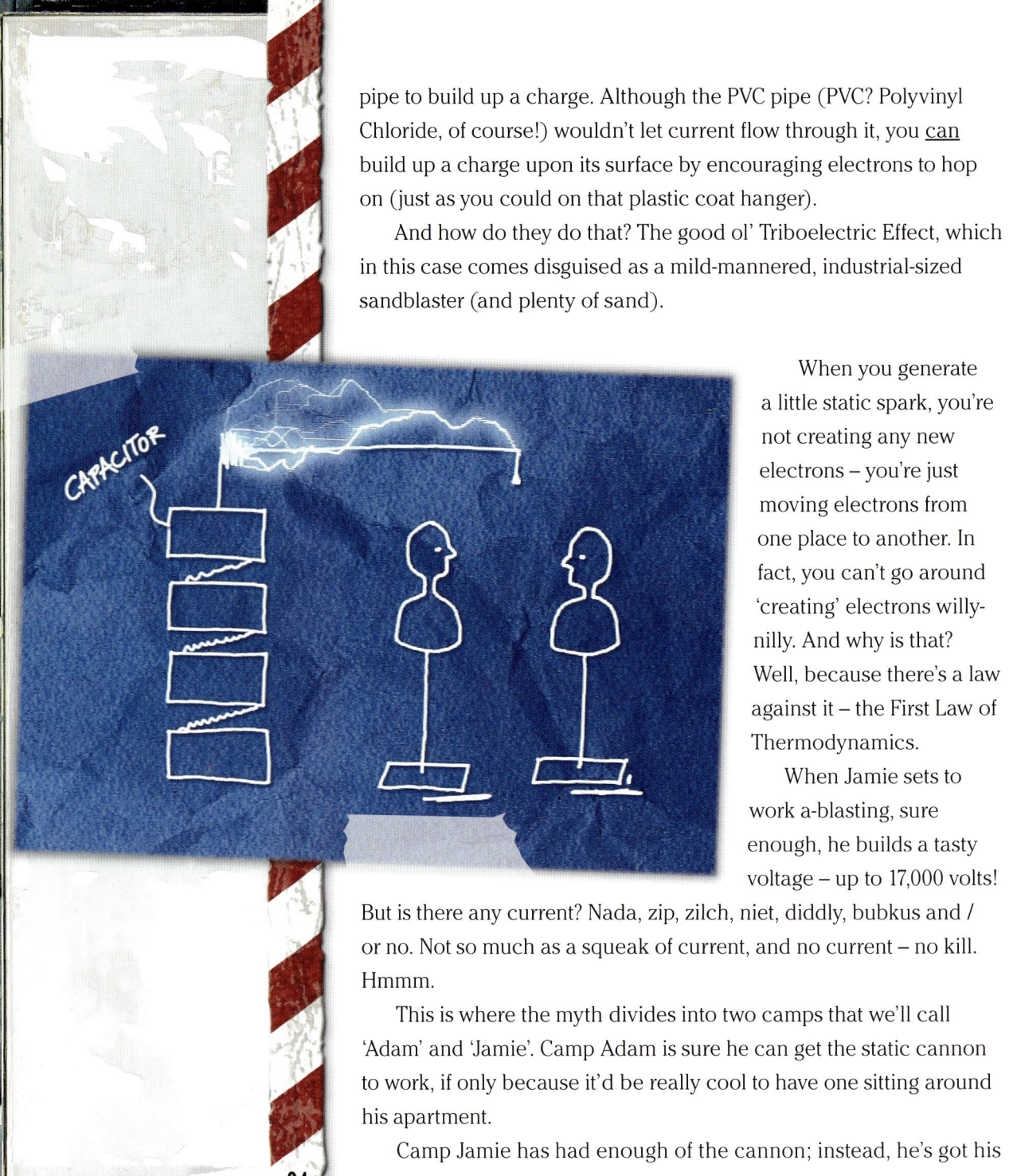

When you generate a little static spark, you're not creating any new electrons – you're just moving electrons from one place to another. In fact, you can't go around 'creating' electrons willy-nilly. And why is that? Well, because there's a law against it – the First Law of Thermodynamics.

When Jamie sets to work a-blasting, sure enough, he builds a tasty voltage – up to 17,000 volts! But is there any current? Nada, zip, zilch, niet, diddly, bubkus and / or no. Not so much as a squeak of current, and no current – no kill. Hmmm.

This is where the myth divides into two camps that we'll call 'Adam' and 'Jamie'. Camp Adam is sure he can get the static cannon to work, if only because it'd be really cool to have one sitting around his apartment.

Camp Jamie has had enough of the cannon; instead, he's got his eye on a gadget he knows will make people's hair stand on end: a Van der Graaff generator. This is a funky device that you'll have seen in a science exhibition at some stage in your life; Jamie reckons he has seen enough of them to make one from scratch.

It's essentially a machine for generating and storing a static charge. The Mythbusters version will use an electric motor to drive a continuous vinyl belt around two pulleys. A 'brush' of bronze mesh acts as an 'electron sprayer', tempting electrons away from the vinyl belt and onto the big hollow metal globe.

These generators can build up hundreds of thousands of volts; industrial-sized versions can build up millions of volts. With that kind of capability, you'd like to think Van der Graaff generators were highly complex machines that require walls of university degrees and endless tinkering to make them work. Right Jamie?

Jamie: *"Well that was a lot easier than I thought it would be. It's actually kind of scary. I have no idea what this is gonna do. Let's find out … go ahead and plug it in. Be ready to unplug it real quick."*

Only seconds later there's zaps aplenty, and Kari's hair is standing to attention. Is it a breakthrough in anti-gravity science? No. With her hand on the dome and her feet ungrounded, Kari's hair is acquiring a healthy sheen of electrons with nowhere to go. Because each hair has the same negative charge, each hair tries to escape from every other hair – and the best way to get away is to stand up.

All good fun (especially when they're zapping Adam in the mooner), but what about the PVC static cannon? In his spare time between getting zapped, Adam has made a close study of the myth. He identifies the fact that the PVC pipe the builder supposedly sandblasted to his doom was painted.

What if it was painted with a metallic paint? If a metallic paint covered both the outside and inside of the pipe, it might act as a capacitor and store a charge!

A capacitor?? A capacitor is basically a device that can store up an electric charge between two pieces of metal – an electricity sandwich if you will (though you will seldom hear it called this in electrical engineering circles). The entry-level model was invented in 1745 in Holland.

Adam: *"So this is called a Leyden jar and it's actually just Tupperware with foil on the inside and foil on the outside. And it's an early capacitor, which is basically an energy storage device."*

Adam starts in with the ol' Triboelectric treatment and the Leyden jar works a treat, sandwiching 9,500 volts of happy electrons; now for the big daddy version. Adam slaps some chrome paint around the inside and outside of one end of the big PVC pipe, hoping that it will act like the foil in the Leyden jar, and then sets to work with the sandblaster.

Sure enough, they build voltage – but this time they actually register a current! All of 20 milliamps! Let's party!

Sadly, no burly builder is going to be permanently floored by this kind of DC zap (not sad for the builder – but you know what I mean), and the Static Cannon is finally tossed in the 'busted' pile.

But do not despair, zap-hungry reader, there's bigger static zaps out there … for example, imagine a static zap that was 100 million volts at a staggering 100,000 amps. Now imagine these zaps zapping the planet 100 times every second. Now stop using your imagination because we're talking about a very real phenomenon called lightning.

The Zeroth Law of Thermodynamics isn't a joke, and isn't much to do with this chapter, but we have to tell you how its crazy name came about. The Zeroth Law establishes that if two bodies are in thermal equilibrium with a third, then they are also in equilibrium with each other – and somehow this wasn't established until midway through the 20th century.

However, by this time the First, Second and Third Laws of Thermodynamics were firmly established in their rankings, and everyone was loathe to move them. But Zeroth Law was deemed such an important principle that it was bumped up from 4th into 0th place. Funny how the world works.

LIGHTNING – THE BIGGEST ZAP ON EARTH!

From rainstorms to cyclones to whirlpools, fans will know that the Mythbusters tackle the many myths surrounding our planet's natural phenomena with fearless determination but mucho respecto.

So you'll understand that Adam and Jamie's antennae began to twitch when they uncovered internet claims that bolts of lightning were bullying people with tongue-piercings.

Adam: *"The Myth is that wearing a tongue stud increases your chances of getting struck by lightning."*
Jamie: *"Like a Lightning rod? … Sounds kinda silly."*

Jamie and Adam busted this one by planting two ballistics gel heads on identical stands. They fitted one gel head with a stainless steel lip stud then zapped both of them with 'mini-lightning' produced at a local power station.

A few dozen zaps later the Mythbusters concluded that while there was a chance metal jewellery <u>might</u> make you more prone to being hit, the lightning somehow wasn't attracted to a tongue stud itself. It took Adam piercing the gel head with a metal door handle to get that result!

Lightning is such a **wow** phenomenon it's no wonder dogs hide under the sofa to get away from it (especially after all that nose static work). Compared to Floridian golfers or Cuban fruit pickers, the dogs are the smart ones. Lightning is so powerful and so potentially lethal that human beings haven't even come close to replicating it, in the lab or anywhere else.

A lightning bolt can be five times hotter than the surface of the sun – nearly 30,000 degrees celsius! Lightning from a single decent thunderstorm could probably power the whole USA for nearly half an hour. Lightning can strike 10 kilometres from the cloud it came from (the classic 'bolt from the blue') and, yes, lightning can strike the same place twice (the Empire State Building has been struck something like 7000 times).

So, you have been warned! Don't sneak out in a thunderstorm wearing a suit of armour and carrying a set of golf clubs tied to a kite attached to a set of keys (more on Ben 'Jackass' Franklin later).

If you get scared thinking about the earth's lightning, don't go wandering around the solar system looking for a safe haven. Jupiter boasts lightning 10 times as strong as ours, but even that is nothing when compared to its be-ringed neighbour, Saturn. NASA's Cassini spacecraft found evidence that lightning on Saturn might be a million times more powerful than Earth lightning! They might even cause streaks in those famous rings!

Okay: so is it a stupid question to ask how such a hot, powerful and deadly phenomenon like lightning is created? No, it's not a stupid question at all, because (and this may sting a little) even the scientists haven't agreed 100 per cent on how lightning works.

Yes, they do realise that lightning is rather a hot, powerful and deadly phenomenon to not really understand, but as most scientists are found in universities (rather than farms in Cuba or golf courses in Florida), there are other things with better funding that they prefer to make their names in.

However, a scientist can tell you that lightning is generated out of electrically-charged storm systems. Prod the same scientist for more details and she might tell you to go get your own PhD and find out yourself.

[9] There are other theories as to how lightning is formed – including a popular idea that the sun charges up the outer atmosphere of the planet with its solar wind. If you want to know more about this, feel free to pick up one o' them PhDs.

[10] You can search the internet for the design of a kickass capacitor that will literally kick your ass - if you're the sort of person foolish enough to bring your ass into contact with the thing at the right time. Who knows? You might just be that person. You've obviously spending a lot of time on the internet … but then, so do we. Maybe WE'RE that kind of person (people?), in which case we should stop shrinking from our true self, get out there, make an ass kicking capacitor, drop our strides and live up to our deepest needs and desires.

[11] For much, much more about lightning, electricity and the human body, reach for that copy of *Rock the Body.*

Or she might explain lightning this way [9]: inside big clouds there's lots of water droplets evaporating (turning into water vapour, 'vaporizing' if you like) and condensing (the opposite of evaporating). The freezing temperatures inside the cloud has a similar effect – essentially the tiny droplets get knocked about and rubbed around, making lots of contact, and some of those frisky electrons jump off some droplets and onto others.

These droplets with extra electrons gather at the bottom of the cloud, giving it a negative charge. The droplets that have lost electrons gather at the top of the cloud, giving it a positive charge.

You can guess what's going to happen, but before we let the world's biggest static spark out of the bag there's some more science to explain. Hang in there.

These charged-up clouds are now acting like capacitors, which you'll recall are devices that store up a charge between two pieces of metal. Capacitors in your household electronics might only be millimetres across [10], but in the case of lightning clouds, they might easily be kilometres across – good news for scientists, bad news for golfers.

This process creates 6000 bolts of lightning every minute of every day of every year, yet somehow you're still more likely to be killed driving a late model sedan on a short journey along a wet road [11]. It does sound like a lot of finicky meteorology needs to take place before a single bolt of lightning is produced, but there's more.

When a bolt of lightning is ready to explode out of a cloud, it starts by forming a channel through the air from the cloud towards the ground. Freakily, as this channel approaches the ground, another channel forms from the top of a tree, or the Empire State Building, or the one iron you're holding, and heads skywards.

These channels (also called 'stepped leaders' for very good reasons that we won't go into) are low current and almost invisible, but when they meet, they close the circuit, and with a "fizz … bang … crack", you've got yourself a bolt of lightning (and one less golfer), all in a few fractions of a second.

When you hear Jamie or Adam say 'complete the circuit' or 'close the circuit' they're not talking about a lap of the local oval or their M5 security system. They're talking about an conductive path that allows electrical charge to flow like traffic at three in the morning (gloriously free and easy).

So: static sparks can jump through the air, whether it's a few millimetres or a few kilometres. However, as the Mythbusters discover when they face up to the ghost of Benjamin Franklin, they travel through some materials better than others.

FRANKLIN'S KITE

If you grew up in the USA (or pretty much anywhere else) you'll have heard of the remarkable 18th century inventor and scientist Benjamin Franklin. You may even have heard of his groundbreaking experiment with a kite, a key and a bolt of lightning.

Tory: *"What have you got there, a little doll?"*
Grant: *"No, it's called an action figure and it's Ben Franklin."*

To prove Franklin's suspicion that lightning was a form of electricity, the myth suggests the great man tied a key to his kite string then flew that kite out his window during a thunderstorm. When a bolt of lightning struck the kite, the lightning flowed down the kite string, gathered in the key, and gave Franklin a headline-grabbing, history-making, paradigm-shifting zap.

But did it really happen that way? Could lightning really travel down a kite string? Could it gather in a key? And why? [12]

Get ready to enter the world of conductors, resistors and insulators.

We've already discovered that it doesn't take too much effort for electricity to travel through the air, even just the short distance from

[12] The Mythbusters also cast a bit of a shadow on the myth by demonstrating that if the experiment really had taken place as it supposedly did (in a thunderstorm), it would have been the last thing Ben Franklin ever performed. Back to the main text for more, and thence your well-thumbed copy of **Rock the Body** for the gritty realities of electrocution.

your finger to your dog's nose. However, there are some materials through which electricity really likes to travel, and some materials that put up more of a challenge.

Materials can be conductors or insulators, depending on the level of resistance they put up to the flow of electricity. It's all to do with those frisky electrons; in some materials – the ones that are good conductors – the electrons are SO frisky that they actually dance around in the space between the atoms looking for action (the shameless hussies).

The good-conducting materials with these 'free electrons' tend to be metals – and metals like silver, copper and gold are some of the best conductors around because they have the most free electrons, and offer the least resistance for an electric charge that's looking for fast flowing fun.

On the other side of the resistance equation are the dour, no fun (dare we say Calvinist?) insulators. Insulators keep all their electrons under tight rein, and thank goodness they do. It's the intolerance for freewheeling electrons of insulating materials like rubber, ceramic and glass that keep killer zaps out of harms way.

Free electrons move through a conductor much like kids in a really, really crowded water slide. When one kid jumps in at the top, that pushes a kid out at the bottom; each kid travels relatively slowly, the effect can be close to instantaneous. And when we're talking about the flow of electric charge, we're talking about the speed of light; well over a thousand million kilometres an hour (it'd be a steep old waterslide that could match that).

But what about something like string? Where does it fit into the resistance picture? Well, as the Mythbusters discover with Franklin's Kite, string can conduct an electric charge — especially when its been hanging from a kite in the rain.

FRANKLIN RESURRECTED!

The myth that Franklin was a lightning-chasing 18th century Jackass bastardises what was in fact a bold demonstration of the fundamental nature of electricity. The Mythbusters replicated and confirmed the real experiment by flying a kite in clear conditions, discovering a measurable static charge was generated on the kite simply through its contact with the air — let's see Johnny Knoxville think up something like that!

Water is a conducting curiosity. Pure water has many insulating properties, but when your bucket of pure water is mucked up with some impurities (especially dissolved salts [13]) suddenly there are free electrons again and the water can conduct a charge. So what might happen if your house was struck by lightning and you were standing in the shower? Jamie and Adam's lightning reactions continued with *Phone in a Thunderstorm*.

[13] 'Dissolved salts' sounds so much more science-y than 'wee in a bucket of water'. Again, for more on the conducting potential of wee-wee, reach for **Rock the Body** (aren't you glad you bought it?).

[14] If you're interested in learning more, get yourself a short-sleeved shirt, a pocket protector and a job in architecture, structural engineering or local government.

PHONE IN A THUNDERSTORM

"Jamie, I know the phone wasn't invented when you were born, but you know, about the time you were growing up, were you ever told to stay off the phone in a lightning storm?" – Adam

"Yes, actually. We were told to stay off the phone, out of the shower, don't go under a tree, don't go out in the middle of a field with a large metal pole." – Jamie

"Yeah, that would be bad." – Adam

Oh boy, oh boy, oh boy… this is the one where the Mythbusters zap a whole house! Okay, now let's all just calm down and act like good little scientists. What's the myth? What did they do? How did it work?

As you'll have noted from the previous pages, lightning is big, bad and dangerous to know. It's not the kind of natural phenomenon you want to bring home to meet your parents, but that's exactly what Adam and Jamie do in *Phone in a Thunderstorm*.

And the answer to the question: "If lightning strikes my house, can I get hurt if I'm on the phone / in the shower?" turns out to be 'yes' – <u>if</u> your house has a few simple wiring deficiencies.

What kind of simple deficiencies are we talking about?

It starts like this; your house (or 'dwelling', as local statutory authorities probably call it) is most likely built 'to code' or under a pretty rigorous building code of practice. This building code is a set of rules that covers everything from where your sewage runs to how your fire stairs are built, as well as the standards of your wiring [14].

Along with other elements of the code, building code standard wiring is all about making your house as safe as possible from the dangers of electricity, so that every piece of toast you make doesn't mean a call to emergency services.

There's a whole bunch of wires involved in the electrical wiring of your house (no

kidding) and most of these are plenty busy carrying power to your outlets and light fittings. But there are a few piddling little wires that just end up either attached to a long pole that's been pounded into the ground, or to the foundations of the building, or to the main water inlet pipe.

These are ground (or 'earth') wires, and they are actually quite important … ie, they protect you from the stray zaps that would otherwise frustrate you from sucking in your next breath. The chief purpose of the ground wire is to safely conduct away charge into the earth in the case of an accident. This might happen if an appliance – say, that mauve '50s vacuum cleaner you inherited from Gran – has a short circuit, like a loose wire, that conducts electricity to the slightly rusty chrome case, and makes it 'live' with electric current.

However, because the case of an electrical appliance is usually cunningly connected to its own ground wire, the current is directed along this wire, into your home's ground wire thence away to be safely 'grounded' – far, far away from your precious internal organs.

The other reason you want the wiring of your house to have a solid ground connection is in case a bolt of lightning comes to call.

A short circuit occurs when an electric charge is allowed to flow along a different path to the one intended; perhaps from the wall outlet through the vacuum cleaner, then into that loose wire, from there through the case to your left hand, through your heart down to the big toe on your right foot (rather than from the outlet to the motor as the manufacturer intended). This is called electrocution, and sometimes it's the last thing you experience.

Jamie and Adam test the myth that lightning can zap you through your phone at a purpose-built lightning facility owned by Pacific Gas and Electric. Their tests reveal that a good solid ground wire <u>might</u> save you from being fried while on the phone, but the Mythbusters reason that the puny power the facility can generate (700,000 volts) isn't giving them fair comparison with the 100,000,000 of lightning.

So what does Adam do? He cuts the ground wire.

Next test? Bingo 'bango' – they fry 'Chip' the chatty ballistics gel dummy (and start a nice little fire to boot).

Result? Adam and Jamie won't be making any calls from this ol' house when the lightning walks about.

We know that electricity can flow through the air, but consider that each 10,000 volts adds about one centimetre to the distance it can travel. Now look at the length of a lightning bolt … Getting scared yet, Tiger? Yah, <u>sure</u> you're not…

Next up in the myth is whether you'll get a lightning-sized zap in the shower. Now, think back – we mentioned that a ground wire is sometimes attached to the water inlet pipes. Well, not only that, but a showerhead AND the plughole / drain are also grounded.

In fact, every metal plumbing fixture in a home should really be grounded. Why? Because if an uninsulated live wire was to come to rest gently on a plumbing fixture, then there's a good chance every tap, drain, showerhead could be charged up and ready to zap – unless they're grounded.

Unfortunately this oh-so-sensible safety valve can come back to bite you when there's a lightning strike nearby. The lightning is also searching for the quickest way to the ground, and – golly – it might just take the route via the shower head, your shoulder, heart, and out through the foot that's standing on the (grounded) drain.

And so powerful is a lightning zap it doesn't even need to zap you through the water coming out of the shower, it can just arc through the air. But don't despair, we'll be looking at more myths with both water and zap next, when Adam and Jamie relieve the tension on the third Rail…

When watching 'Phone in a Thunderstorm' did you notice how the locked-off camera lost focus every time the current was zapped into the house? That was due to the electromagnetic pulse (also given off by a nuclear blast) produced by the zap – and the camera was 10 metres away! AND those zaps weren't even a fraction of the power of a lightning bolt! Sheesh … that electromagnetism is crazy stuff!

TINKLING WITH THE THIRD RAIL

Railway workers share a legend that speaks of an Irishman called Joseph Patrick O'Malley. One night after a few ales this stout yeoman was walking home past an electrified railway yard when he felt the call of nature …

Adam: "*Oh, this is the one the guy found himself late at night in a train yard, peed on the third rail, the electrified rail, got electrocuted and died.*"
Jamie: "*That's the one.*"

Zapped through your wee-wee? Ouch. It's a good story, but what factors might lead to it actually happening? And … uh … what's this third rail? The electrified third rail sits close to the two main rails of a railway track. A metal plate, or 'shoe' slides along the rail, conducting power to the engine. [15]

[15] Yes, we realise there are many systems in use in different railways around the world, but this is all we're prepared to say about the working parts of trains. Why? Because we know that trainspotters and Mythbusters go together like trainspotters and anoraks; if we carry on any further, we'll get stuff wrong BECAUSE WE COULD CARE LESS. Then you'll be forced to fire off emails and petitions to the publisher demanding levels of detail that are inappropriate for a book of this kind. So if your idea of a good time is waiting in your Robin Reliant to see if the 3.37 from Horsham has the new split-bogie trans-circuit 'horse-head' hotbox detectors, then give us a break will you?

The third rail carries relatively low voltage electricity, in the region of 650-1000 volts. The overhead wires used in other train systems can deliver much more power (25,000V to 50,000V) because, unlike the third rail, they don't carry the risk of arcing to the main rails.

But this lower voltage means that to get sufficient power to the train, it needs to come at a much higher current. As we know – it's the current that kills (although peeing against an overhead wire won't be the best way to get a letter from the Queen either).

We promised you the skinny on Alternating Current – well, as opposed to direct current, which flows in one direction only, alternating current alternates – it flows back and forth. How and why? We'll, back in the good old days two inventors were vying to zap the planet with a new power called 'electricity'. They were Thomas Edison and Nikola Tesla, and they started a kind of war over how this electricity should be delivered.

Tesla – who was a clever fellow with all sorts of maths behind him – concluded that if you could make the current zap one way, then the other ('alternately') it would be more efficient, because the wires wouldn't heat up and burn the power before it got to where it was going. Edison (who was off sick when they did maths at school) didn't get it, and championed the direct current, which was easier to understand. In the end Tesla won, but nevertheless died alone and in poverty, while Edison died filthy rich surrounded by drooling supplicants who even captured his last breath in a test tube. Funny thing, life...

Railways sometimes use AC, sometimes DC. Jamie and Adam need matching third rail voltage for their experiment. To do so they set up a rig that will use a transformer from a fluorescent light. Of course they do – who wouldn't?

Okay, now you're in transformer territory, and we're not talking robotic action figures with more than meets the eye. Transformers are nifty devices that can 'transform' electricity into higher or lower voltages (or in fact currents) as required for an application; in this case, the rig for a possibly fatal wee wee-zapping experiment.

A transformer is kind of like gearbox in your car, which is there to turn the power from the engine into either more speed (in a higher gear), or more grunt (in a lower gear).

In the case of electricity, a 'step up' transformer trades current for voltage, and a 'step down' transformer – you guessed it – trades voltage for current. The overall power remains the same (we're not breaking any of those fundamental laws here) because of the cosy relationship between voltage, current and power.

Large fluorescent lamps that operate on 120V mains (such as in the USA) need a 'step up' transformer to give them enough voltage to zap the mercury vapour inside the tube into action. Thought you'd like to know …

Okay, okay, okay, we've spent quite some time on more electricity factoids and devices (and there's more to come), but there's something we've so far avoided. Perhaps we're shy, maybe we're embarrassed, but we know and you know that it wasn't the world of Alternating Current and transformers that made you flick to this page.

It was the human wastewater we call urine, piss, pee or wee wee – and it's taking the stand right now. If you paid close attention you would already have learnt that water diluted by salts was a fine conductor of electricity.

Well, that's urine. But urine, and all conductors or insulators, have a resistance that can be measures in Ohms.

[16] Alright, so the ancient and sacred Hindu symbol of the infinite Brahman and the entire Universe is spelled Aum. But hey - it sounds the same! Furthermore, we reckon that if you attained spiritual oneness with all things (perhaps by weeing on something electrical) you could probably think up a really good 'hey man' connection between Ohms and aums.

Nineteenth century scientist Georg Ohm's work on the intimate relationship between current, voltage and resistance was deemed sufficiently hair-raising for him to become a member of a very exclusive club. Along with Alessandro Volta, Andre-Marie Ampere and James Watt, Ohm's name been attached to the units of measurement of a major electrical phenomenon, and simultaneously all have become much-hated figures in examination rooms around the world.

So what on Earth is an Ohm, and where can I get one?

Ohms have nothing to do with meditation [16] and everything to do with how a material reacts to a good, hard zap. But there's nothing kinky about an ohm (okay, well maybe there is – for more flick back a few pages to Franklin's Kite).

Signified by Ω, the ohm is the unit of resistance – which was (if you recall) the way that a device or material resists the flow of current. Adam and Jamie measure the resistance of urine with an Ohmmeter (of course) and calculate the resistance of their experimental widdle should be 10,000 ohms.

The relationship between the resistance of a material and the current it can deliver is so intrinsic to the kind of zap you'll get through it, that there's a law relating them (yes, another one).

This one is Ohm's Law, and it means that if 650V came at you through a stream of urine with a resistance of 10,000 ohms, you would get a zap in your bits of 65 milliamp – enough to take care of you forever.

This cosy relationship between volts, amps and ohms means rather conveniently that when you start dropping electrical appliances into your bathtub, you can work out one as long as you know the other two.

When you start throwing <u>what</u> into the <u>what</u>?

Adam and Jamie busted the myth of the third rail by demonstrating that urine could not conduct electricity because it breaks up into droplets before hitting the ground. Why does urine (and other liquid) do this? It's all to do with surface tension and air resistance – but as neither has much to do with electricity, we'll leave the deeper explanation to another book in the series. Suffice to say that for Adam to get a shock in the *Third Rail Revisit*, Adam had to urinate on an electric fence from only 10cms away; oh yes, it was quite a day.

Adam: *"Can I point out that I've already been painted with gold paint and had an anal probe today?"*
Jamie: *"You're doing a wonderful job."*

APPLIANCES IN THE BATH

Before we start, please take note of the following special warning: no matter what Adam and Jamie say or do in this episode, you should never blow-dry your hair while taking a bath. [17]

Jamie: *"What's with the bathrobe?"*
Adam: *"Well Jamie, this time we are attacking one of the classic Hollywood movie myths, which is simply, if you threw an electrical appliance into my bath, I'd get electrocuted."*
Jamie: *"Let's go."*

Sadly, because of the Mythbusters' draconian insurance policy, Adam was removed from the bath at the last minute and replaced with a ballistics gel dummy.

The dummy has pretty much the same electrical resistance as Adam would if he's all wet and wrinkly in the bath – around 1000

[17] Oh, seems obvious does it? Let me tell you that when you learn that Francis Bacon, the greatest mind of the 16th century, died after spending a day standing around in the freezing cold while he stuffed snow into a chicken to test his theory that this was a way he could preserve meat, you begin to fear for the fate of all humans everywhere.

ohms. Thus if we drop something in the bath that's drawing on the mains power (120V in the USA) then a rough and ready application of Ohm's Law might give a current of …

Wait for it … it's Ohm's Law guys, the previous page … come on, if you're in that bath you could be dead already!

Okay, while you thrash around with steam coming out of your ears (and elsewhere) let us tell you that the current is the voltage divided by the resistance. 120 milliamps; definitely a killer current.

Jamie and Adam hit their dummy with a toaster and an iron for maximum points and a confirmed kill every time. In fact, there was only one appliance that wouldn't come to the electrocution party.

The modern hair dryer.

Most modern bathroom appliances are fitted with what electrical engineers might have called an Anti-Death Switch, but otherwise called a Ground Fault Interrupter, or a Ground Fault Circuit Interrupter, or a Residual Current Circuit Breaker, or a Appliance Leakage Current Interrupter (or even an Interruptor de Circuito con Conexiòn a Tierra).

Boring! Yet the thing itself is a very, very useful little device should you be taking a bath while blow-drying your hair (which you shouldn't – see earlier warning).

An Anti-Death Switch (we prefer our terminology – it might catch on) uses a transformer to detect the leakage of more than about 5 milliamps of current from the appliance. If a current 'leak' occurs (perhaps because, oh, we don't know, Jamie and Adam just chucked a hair dryer in the bath tub) the Anti-Death Switch takes only 25 milliseconds to shut off the juice – which is faster than it takes for that juice to cook your internal organs.

The Anti-Death Switch is not dissimilar in function to another simple safety device, and in the next myth we learn about a couple of 'good ol' boys' that learnt the hard way about how they work. We'll be a little easier on you …

There was another key revelation in Appliances in the Bath, and it was this – additives to the water had a huge effect on the resistance of the water. Bubble bath made the water more resistant, urine and Epsom Salts made it less resistant, and hence more dangerous when you drop in your hairdryer. Well, we might have guessed that, but would you have guessed that dropping that same hairdryer closer to the grounded drain would mean less current through your body, because the distance 'to ground' was shorter? Oh, you would have, would you? Well, good for you.

FROG GIGGIN'

"Okay Bubba, what have we got?" – Jamie

 "Well Thurston, we've got a good one today. We've got frog giggin." – Adam

 Let's say that once again.

" … We've got frog giggin'." – Adam

[18] Frog giggin' (or gigging) means hunting frogs, usually with a spear or pitchfork along with a good deal of insensitivity to the aesthetic and environmental value of frogs. However, such practises are as old as humanity. Aesop even wrote a fable about boys who killed frogs for sport – but he didn't think of such a good ending as we've got in this myth. Ironically, a 'gig' is also a slob who looks weird and annoys people by being both boorish and stupid (perhaps by frog giggin').

[19] We can feel the twitching of the gun clubs from here, so YES when we say 'bullet' we really mean 'round' or 'cartridge'. The bullet is really just that part of the cartridge that gets fired out the end of the gun. Oh, all right – when we say 'gun' we really mean 'rifle' because a 'gun' is a piece of artillery of the kind found on a ship or a tank. If you're still twitchy, just remember – we will publish a whole book dedicated to Mythbusters weaponry, and another whole book dedicated to Mythbusters explosions, but ONLY if we live to see another dawn. Deal? That's the story.

Could 'Frog Giggin' be the most ridiculously named myth that's been up for Bustin'? It would certainly rank as one of the most painful for those originally involved – but only <u>if</u> the myth could be confirmed by those two mockers of moonshine, those filleters of flummery, the excoriaters of exaggeration, Adam and Jamie.

In a nutshell (or an old-style pickup truck to be more precise) a couple of good ol' boys were about to return home from a pleasant evening's frog giggin' [18] when suddenly their headlights fail. Reckoning they'll need them headlights to get home in one piece, they put safety first and replace the busted fuse with a .22 calibre bullet. Hmmm. But, lo and behold, the lights go on. Problem solved, right?

Sure, until 20 miles down the road when the bullet explodes and hits the driver such that he'll never again take part in a 'third rail' experiment again, if you know what we mean.

Karmic justice? Rank stupidity? Take your pick, but first, we all know what a fuse does – yes?

… *we're getting more than a little silence here.*

Okay, a fuse is an easily replaceable conductor (often as simple as a bit of wire, in a car sometimes encased in a glass tube – hell, there are many kinds) that will heat up, melt and thus break its electrical connection at a certain temperature.

The temperature at which the fuse breaks represents the amount of electric current above which the more difficult to replace components of the electrical system (your car or home electric wiring for example) would have suffered an expensive and possibly dangerous meltdown.

In essence: the fuse is there to take one for the team.

So far so good, but why would a bullet [19] work as a fuse? And why would it explode? Well, the answer to the first question is: It doesn't work as a fuse.

Sure, a bullet – being made of brass and lead – will conduct electricity. But pretty much the key to what makes a fuse a fuse is that it will fail at a given temperature.

Now a bullet will do this too – and indeed did in the experiment, by resisting the current and thus getting itself all hot; so hot the gunpowder-or-cordite propellent inside the round was ignited and thus the bullet was fired (or 'the round was discharged' for those who read and still feel twitchy).

The picky among you will simply point out that if a fuse is there to fail at a certain temperature, and a bullet (okay, *round*) fails at a given temperature, then what's the difference? We reply, if you can't see the difference, then by all means stock up on .22 rounds next time you're going on a driving holiday.

Back in the real world, it took Jamie and Adam some searching to come up with just the right kind of pickup kitted with just the right kind of fuse box aimed at just the right part of the driver's anatomy. Now this is indeed a lot of coincidences, so it obviously took a heady mixture of rank stupidity and bad luck to happen in real life. We reckon that's what happens when you mess with frogs. [20]

Okay, so we promise not to mess with frogs, but let's wind it back a notch and ask 'Why does electricity generate heat?' (Well, enough heat to set off a bunch of gunpowder?). To answer this apparently simple question, you're going to have to step up and admit that you don't know how fire works either. Go on, be honest – Kari, Grant and Tory were man enough!

FIRESTARTER – TAKE ONE

When Grant, Tory and Kari wanted to prove that fire wasn't a myth, it seemed absurd (and utterly, completely, moronically obvious). Yet these three apparently switched-on human beings quickly found out that they were damn lucky not to be born during the last ice age.

Seems there's more to fire than meets the eye.

Kari: *"The very hardest part about all this is … we know you can make fire while rubbing two sticks together. It's been proven over and over. It's just that … we really suck at it."*

Tory: *"You know if you don't get this lit, we're going to vote you off the island."*

[20] The smart word is that the other famous frog myth (you know the one that says you can boil a frog alive just by slowly heating up the water in a pan you've sat him in?) is also wrong. We say again – don't mess with Kermit!

Here's the link to the previous chapter – the question 'why does electricity generate heat?' has the same answer as 'why can you start a fire by rubbing two sticks together?'

And the answer is … friction.

Remember when we talked about resistance? Where all those frisky electrons dash their way through a conductor like buckets in a brigade, or perhaps marbles through a tube, or kids down a slip-and-slide (we're pushing it there, but it kind of works)? You remember that there's bags of space inside and between the atoms of any material?

Well, there is, but not so much that the free electrons have it all their own way. It's not like a school hallway during summer holidays. Depending on the material the electrons pass through, they meet resistance (flick back to 'Tinkling With the Third Rail' for a reminder); in other words, they bump up against the other stuff inside the conductor.

This 'bumping' (not a scientific term) that the electrons experience creates friction, which can generate heat as well as light.

Light bulbs work because they send electric charge through a material that offers resistance – materials like tungsten (which make up the 'filament' of your standard light bulb) specialise in getting so hot when they have a charge dashing through them that they glow particularly white hot, and 'Ta dah! Let there be light!'

Older light bulbs like the one Grant and Kari visited in Livermore, California, have a different kind of filament that doesn't get as hot, or glow as bright, but lasts longer (in that case 106 years and counting). Friction is pretty cool – although to be more accurate it's pretty hot.

Friction happens whenever materials are in contact and transfer some kinetic energy; that's energy that you gain as a result of motion. Even an object falling through air experiences friction between itself and the air – just look at a shooting star [21]. The more kinetic energy you can 'impart' to a material (that is a fancy science term – remember it!), the more that energy is turned into heat (via interaction at an atomic and even subatomic level), and the more likely you are to approach the material's ignition temperature.

Ignition – now we're getting somewhere.

I know, you're still in the dark a bit – we can tell. Don't be embarrassed, the question 'what is fire' had the Greeks stumped (and they invented democracy … sort of).

The Greeks thought fire was a state of matter (along with earth, wind and water). In fact, fire is a chemical reaction in process. It requires particular ingredients to exist; heat, oxygen and some kind of fuel. Freakily, the chemical process that is fire is much the same process that makes iron rust. Oh yes – and before you say 'but rusting metal

[21] We all know that a shooting star is really a bit of stuff (rocky meteorite, piece of a space station, a tiny alien spaceship that came out of hyperdrive too soon) that falls into the air-laden atmosphere of Earth and burns to nothing (unless you're very unlucky) because of the heat it creates through friction between it and the air, which is very much more than the 'nothing' it seems to be … so then why am I telling you? Get back to the main text!

isn't on fire!' we say to you 'No, but it does increase in temperature!' (by about one degree Fahrenheit – or a fraction of a degree Celsius).

This rusting-or-burning process (as a sciencitian you would call it 'oxidisation') means that atoms of oxygen combine with atoms of hydrogen and carbon to form water and carbon dioxide.

The other by-product of the reaction is heat.

Now, some materials oxidise really fast, so fast that the heat can't be released fast enough, and this means one thing – fire! (again, as a good sciencitian you'll be calling it 'combustion')

Fire is what chemists think of as 'excitingly complex' because firstly there are literally a hundred chemical reactions in even the simplest fire, and secondly chemists don't get out much.

One of the most exciting things when a material is on fire is that it is undergoing what's called a branching chain reaction. As the fire combines oxygen with hydrogen and other atoms, more atoms are released from the burning material that combine with more oxygen, that leads to more heat, and more atoms are released …

See what we mean by a branching chain reaction? If you want to see this in action, go to a nice, dry, bushy place like South Eastern Australia any time between October and March.

But we're not done with the **wowness** of fire. Get this: paper doesn't burn. Uh, huh. And neither does petrol (gasoline).

Wass-at you say?

Okay, it's a bit of a clever-clever thing, but the paper itself doesn't burn. What happens when paper is 'on fire' is that the heat of the fire causes the paper to vaporise into gases, which then burn very happily indeed thank you very much. Same with petrol.

Boy, is there more? You bet. A flame can be invisible.

Go on, you're having a lend.

Seriously, the reason that a candle flame (say) glows orangey-red isn't because of something intrinsic to the nature of a flame. It's because there are little atoms of carbon that the fire (chemical reaction) doesn't completely combine with the oxygen in the combustion process. Instead, it floats around and hooks up with other atoms of carbon, and some other atoms of 'whatever', and forms into a molecule. This molecule, mostly carbon, then turns the energy of the flame into light.

If you're thinking that sounds like the same thing the filament in a light bulb does, then bingo, you are right.

Flames are invisible if the fuel is turned into energy in a very efficient way. Ever been watching a Grand Prix on TV, and seen one of the pitmen jump away from the fuel pump he's handling and throw himself to the ground, screaming in agony, for no apparent reason? He's probably just caught fire, because the kinds of fuel Formula One cars use can burn without a flame. Scary? You bet (sometimes those invisible flames hide under your bed too ...)

If this stuff on friction and fire seems to be tugging us a little bit away from zaps, then (A) we'll be the ones who do the links <u>thank-you very much</u>, (B) we like fire and friction, (C) all things are connected in the greater scheme of ... things, <u>man</u>, and (D) if you keep reading we'll tell you about how you can start a fire with a battery, and IF IT'S ALL THE SAME TO YOU get stuck into some serious facts about how zaps can be generated. Okay?

After hours – days! – of failed attempts, Grant, Kari and Tory called in a survival expert to make their 'firebow' work – which he did in about 10 minutes flat. Result? Fire is not a myth (hold the front page for that one) and Grant, Kari and Tory would have been three very cold Neanderthals.

BATTERY-GENERATED ZAPS

Any reasonably observant viewer will recognise that the Mythbusters aren't above sharing a little pain.

Evidence?

M'lud, we present 'Drive Shaft Pole Vault'; Tory auditions for Jackass by trying to jump a lil' red wagon on his mountain bike (result? Ouch).

Furthermore, 'Exploding Jawbreaker'; a superheated jawbreaker explodes in myth-tern Christine's face (result? Ouch).

Finally, 'Baghdad Battery' (a.k.a. 'Ancient Electricity'); Adam is encouraged to grab hold of a reproduction 'Ark of the Covenant' and experiences a zap of biblical proportions (Result? Language so explicit it will never, ever be broadcast).

Adam: *"Ohh %&$#@^!!!!"*
Scottie: *"Did you get the kick in the chest?"*
Adam: *"Ah, yeah, and the head, and the entire body. That was ten thousand volts through my heart!!"*

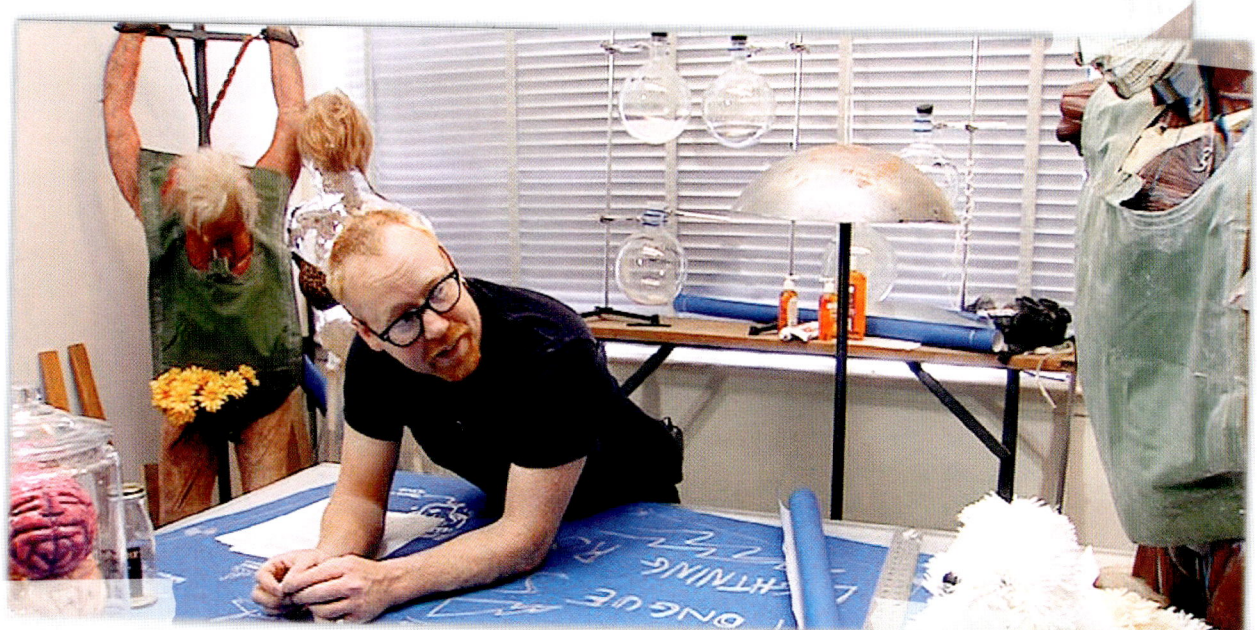

Now, it wasn't the mythical Baghdad Battery that gave him the zap (those crazy kids Kari and Tory hooked in the generator from the electric fence used in the Third Rail Revisit), but, nevertheless, these <u>mucho ye olde</u> batteries are some of the niftier devices the team has plucked from the ancient world.

Batteries generate their power because of a chemical reaction (just as does fire). Now, there are all kinds of batteries, and we're not going to go into detailed descriptions of all of them because ... that's just not how we do things in this book. Get your lithium kicks elsewhere.

The word 'battery' springs from a visual relationship that the early European electricity pioneers saw between their chunky low-zap batteries [22], and an artillery 'battery' lined up to dole out long-range punishment. Imagination? Those guys had it in spades. The ancient batteries the Mythbusters successfully replicated consisted of a terracotta pot, a copper pipe, an iron rod, an acidic liquid and an asphalt cork.

Sounds like the start of a joke? It certainly had a lot of archaeologists very excited. See, these elements were supposedly not put together to make electricity before about 200 years ago when the history books relate that one Alexander Volta (remember him from earlier?) used copper, iron, acid-soaked cloth and a bit of wire to make the 'first' battery (or 'Voltaic Pile' as he modestly called it).

But the question before us now is ... how does this grab-bag of stocking-stuffers (weird stocking [23]) make electricity in the first place?

Basics first – if we're talking electricity, then we're talking about a flow of electrons. If these Baghdad Batteries work, then something is making electrons move around between the iron bar and the copper

[22] Word is that the term 'battery' was in fact first used in relation to a 'battery' of Leyden Jars – which we'll recall are early capacitors used to store electricity. But, that would have made that little piece unnecessarily complicated, so we ignored it.

[23] Weird Santa more like. Who leaves you an iron rod, a copper pipe and a tub of acid for Christmas?

pipe. And the batteries did work; Kari, Scottie and Tory put together a rack of these ancient gadgets and – wired together – they generated a zap of 4.3 volts.

All 4.3 of these volts are thanks to the wonders of electrochemistry, friends. Remember that copper was a good conductor? It was a good conductor because it had lots of free electrons that were looking for action. Iron, however, is a bit further down the list, without so many free electrons.

What if you could convince some of the free electrons from one of these metals to hop on over to the other metal? Well, you'd have an electric current. But how to persuade them to do this? Lollies? Trips to the movies? A well-constructed argument explained with conviction and panache?

What you need is an electrolyte. Yes, yes, yes – just like those expensive drinks you buy when you're pretending to exercise. You know how they taste kind of salty? Kind of tart? Well, that's because they've got salt and acid in them.

Now, those bottled electrolyte drinks are designed for the human body, but an electrolyte solution can be something as simple as lemon juice, or grape juice, or saltwater. Remember how 'water with dissolved salts' was a good conductor? It's back baby, and it's doing something new.

The electrolyte in the Baghdad Battery (in fact grape juice) convinces each metal to pop out a few electrons, as part of a chemical reaction that produces hydrogen. The copper pipe (let's get science-y and call it an electrode) gives up more than the iron rod (likewise, electrode) because it has more free electrons to lose. This means that the copper pipe is now building a positive charge (more so than the iron electrode), and it's just itching to get some electrons from somewhere else to redress the balance.

Well, those electrons aren't going to dance across the electrolyte. But, if you connect the iron and the copper with a wire, the copper electrode gets its wish. It is now a cathode (science talk for an electrode that is gaining electrons) and the iron electrode is an anode (giving electrodes up).

Of course, once Kari, Tory and Scottie realise they've got electricity on their hands, it's not long before they start hurting people with it. First, they grab a qualified acupuncturist to use some ancient electricity to zap Scottie's chakra (and yes, it burned more than a little). Second, they built a reproduction ark of the covenant and connected it to the electric fence generator (and that hurt too).

The rest is history.

Frogs were in trouble again when 18th century Italian electricity pioneer Luigi Galvani discovered he could make a dead frog's leg twitch by touching it with copper and iron rods that were connected with a wire. But frogs took their revenge; it was not Galvani but Volta who turned the concept into the first battery (the Baghdadians not withstanding) and then Mary Shelley, who turned the frog into a fictionally restitched human corpse and made herself a worldwide horror celebrity. Frogs rule.

[24] We're sure that some kind of warning should accompany that step-by-step guide to starting a fire, but then we thought – why? You could go and strike a match and burn down your house/school/parliament just as easily. Not that we're suggesting you do, of course.

Modern batteries use different materials, but the principle is not so vastly different that we can't get away with saying 'it's pretty much the same' and expect just the usual truckload of geek hate-mail.

But just in case, let us instead distract you all with a quick experiment; if you've got a six- or nine-volt battery (with the electrodes on the same side – you know the ones), some old-fashioned steel wool, and a pinch of dry, fluffy tinder, you can replicate the 'firestarter' experiment from episode 48 (just don't burn your place down doing it). Fray up the steel wool and get it touching both electrodes. This 'short circuit' causes a spark that you'll try and catch in a pinch of dry tinder. With a little practise, you'll be sitting in front of a blazing fire in no time (see above warning). [24]

Okay, so you're warming up in front of your post-modern battery-powered fire, chowing down on a roasted leg of some unfortunate slow-moving beastie, when a thought strikes you – we know how fire is made – but … what makes it hot?

Ahhh, little Grasshopper, you are about to enter a World of Rays…

"Ooohh %&$#@! ... that was ten thousand volts through my heart!"*

The Basics of Rays

Chapter 2

The problem with rays (even death rays) is that, depending on how you look at them, they're also waves. This conundrum is compounded by the language we use, which often describes rays (or waves) as waves (or rays) when they might better be describes as rays (or waves) while some rays (or waves) might better be described as waves (or rays).

But let's drop the 'wave particle duality' stuff and get serious about rays. At some stage in your life you may have heard about something called 'the electromagnetic spectrum'. No? Well then, welcome to a startling concept that may change the way you look at everything around you.

Ever seen a rainbow? Ain't they lovely? Well, a rainbow is just a cheeky and colourful sliver of the electromagnetic spectrum. As you gaze lovingly at the wonder of nature that is the rainbow, notice (if you can tear your mind away from the abject beauty of it) that the colours of the rainbow always come in the same order.

The light rays that are refracted by airborne droplets to make a rainbow are divided into different wavelengths (see the problem with rays and waves?).

"Our death ray doesn't seem to be working. I'm standing right in it and I'm not dead yet." – Jamie

Refraction is something that happens to a wave when the medium (a.k.a. stuff) through which the wave is travelling changes. Say a ray of sunlight is zipping through the air (one medium) when it strikes a droplet of water (another medium). Because of the different density of water to air, as the beam bounces around inside the droplet it slows down just a fraction. However (and this is the cool bit), that ray of sunlight actually includes a wide range of different wavelengths, and all these wavelengths slow down at different speeds.

An (almost) accurate analogy might be to imagine athletes sprinting in the different lanes of a running track; they're all running at the same speed and are all in a line when they hit the bend. All of a sudden the one on the inside is ahead, the athlete next to him a little behind and so on. And inside that droplet of water what was once a ray of sunlight has 'dispersed' into the spectrum of colours it contains; the rainbow.

Much as the red and orange and yellow and green, blue and indigo and violet (we know, it's different in the song) are the kinds of light that you and I are most familiar with, they are in fact just a tiny, teeny fraction of a much bigger spectrum of rays (or waves).

This spectrum includes everything from the waves that deliver your favourite TV show to the X-rays that your dentist uses to scold you because you've eaten too much of the chocolate advertised during your favourite TV show [25] (see how the Universe is one big interconnected thingy?) and much, much more. The spectrum also includes the radar waves your airport uses, the infrared rays (or waves) your remote control uses, the microwaves your microwave uses and the ultraviolet rays (waves?) the sun uses to fry your skin at the beach. At the very top end are the gamma rays the aliens use to shoot sci-fi heroes with before they can foil their evil plans.

And all of these rays (or waves) are made up of things called photons.

[25] We really like that analogy and we hope that it helps (and isn't massively misleading in some unforseen way). You could even imagine the athletes in different coloured running suits! Anything is possible inside that head of yours.

Photons are pretty amazing things – as amazing as electrons but for excitingly different reasons. For a start they have no mass. That's right, no mass at all. You might think they were not there at all, except for the fact that they are brimming with significant amounts of energy.

Some of them are VERY significantly full of energy. The photons that make up some rays (or waves) are so full of energy that if they hit you, they start to … well, they start to change you, entering your body and encourage electrons to escape from the atoms that have been chaining them down just forever.

This is not good for your body (which might be why bungee jumping into nuclear accidents or sun tanning in gamma rays are rare lifestyle choices). Such rays have been given the deadly sounding name 'ionising radiation'.

Ionising radiation is not to be taken lightly (no pun intended – cheeky though it is!) and fortunately there's not that many opportunities to come across it – unless you chase nuclear accidents, play with your dentist sister's X-ray machine when she's not looking, or go out into the sun underneath a hole in the ozone layer. Fortunately microwaves and radio waves offer no chance of any ionising, thus science has crafted the term 'non-ionising radiation' to underline its status as gentle, warm and safe.

THE BASICS OF RAYS

A truly deep understanding of photons is the preserve of the very, very clever. People like you and I get in and out of the photon rather more quickly so we can watch television sports and review product catalogues. Some people devote lifetimes to the intricacies of things they can never even hope to see.

But from the perspective of the television and catalogue set, we think the three most vital things you should know about the photon include …

1. Photons make up all the waves and rays on the electromagnetic spectrum (the light from your laptop AND your laptop's Wi-Fi connection AND the invisible rays cooking the DNA of that flawless beach babe / hunk in the photo on your desktop – it's all photons baby).
2. Photons exhibit some of that 'wave-particle duality' that so frustrates scientists; they're the very things that can be either a wave or a particle, depending on their mood (or more particularly, the needs of the observer – we said it was weird).
3. Photons are very, very speedy. In the vacuum of space they travel at – gosh – the speed of light. Because they are light (sometimes).

At the top end of the electromagnetic spectrum are those Gamma rays. The photons in these rays (or waves) are very energetic little fellows, so energetic that the only things that produce them are funky stuff like radioactive decay and the nuclear reactions in deep space.

When Gamma rays are being wavelike, their waves are a very, very short distance apart (that is, they have a very short wavelength), and the waves zip past really quickly (that is, they have a very high frequency). The higher the frequency, and the shorter the wavelength, the more energy in each photon.

So as you travel down the spectrum from the very high energy, high frequency, short wavelength gamma rays, you encounter the different wavelengths and frequencies we have dubbed X-ray, the ultraviolet ray, visible light, infrared rays, microwaves and down to radio waves, of which the extremely low frequency versions can have wavelengths that are hundreds of thousands of kilometres long (the gamma ray, in comparison, has a wavelength as short as one hundred billionth of a meter – if not shorter).

Just to confuse matters further, scientists also describe all of these rays as 'radiation', probably because they involve these energetic photons radiating outwards from somewhere. As you'd expect, when humans realised these rays (or waves) were out there zipping around the place in interestingly fast ways, we began the process of taming them to our will.

Thus, in 'Rays' we'll go into all those funky myths that have to do with radios, radars, microwaves, MRIs and … oh, what was the other thing? Oh, yes.

The Sun.

We'll leave all the other waves (sound, pressure etc), which are termed 'mechanical waves' by people in the know, to the bit of the book actually called 'Waves'. We'll explain more about that then.

Okay, can we get back to the myths, please? Thank-you.

FIRESTARTER TAKE TWO

Close viewers of the series will note that the episode *Firestarter* included several fire-starting experiments that we didn't cover in the prior chapter we suggestively called *Firestarter Take One*.

These experiments will take centre stage as we deepen our discussion of 'rays'; good luck to you all.

Tory: *"The idea is to take a block of ice, rub it until you get the shape of a lens, and then taking the sun's rays and focussing them into your kindling."*

We knew you'd love it – but if you like the 'fire from ice' experiment, then you'll love 'fire from chocolate and soda'!

Grant: *"So the idea is you take a piece of chocolate, rub it on the bottom of your cola can … and after a time …"*
Tory: *"We're going to try to polish this up to a mirror finish."*
Grant: *"Right."*

Polishing things with chocolate? What do the *Mythbusters* do in their spare time to come up with these ideas?

On second thoughts … let's not go there.

Both of these two firestarting techniques rely on a helping hand from a certain stellar object that's not so far away that looking directly at it wouldn't cause problems in the eye department. Different cultures have named it Sol, Helios and … the Sun.

We're not overstating the matter when we say that the Sun is a big, round, hot, important thing. It even has a capital letter [26]. In fact, we could easily get away with saying that if the Sun wasn't out there in the centre of the solar system, simply bursting with joy at being

---

[26] Interestingly, not everything in the solar system gets this treatment. Just read on… yes! You notice that the solar system itself is not capitalised! You see, the Oxford English Dictionary doesn't, nor do other dictionaries and encyclopaedias of repute, although the International Astronomical Union, which claims authority over the naming of the specific bits and pieces of the universe (quite a call) does require the capitalisation of all individual astronomical objects. Is the solar system an individual object? Depends on where you're standing, or how big you are. So if you care, then for goodness sake, throw yourself into this vital debate by contacting the punters at www.iau.org (a friendly bunch who love to get unsolicited opinions from the public).

[27] Because the Earth's orbit is elliptical, this is just the average distance.

[28] Recent discoveries have turned up even clever-cleverer life forms who make a tidy living from chemicals that spew out of the very Earth itself – which is only here because of the Sun's gravitational power, leading to the formation of planet-sized rocks, et cetera, et cetera … so, yes, the Sun is still responsible.

[29] In fact, air resistance makes the parabola less than perfect, but at low speeds you can get away with it. If you were firing a missile at North Korea on the other hand, you might want to get a maths monkey do you a few more calculations (and then a diplomatic monkey to sort out the subsequent crisis).

a massive ball of hot gas, then you and your favourite jeans and your dog and your dog's best friend Gravy (the cute but boisterous Staffy at the local dog park), would not even be a figment of anyone's imagination – because that 'anyone' would have to be an alien who's probably more interested in recalibrating his gamma ray gun than imagining cute but boisterous dogs on non-existent worlds.

In short, without the Sun we would be nothing.

The Sun is important because it does some clever stuff that keeps the whole planet ticking over. Reactions deep inside its core and at its surface (don't ask me to explain these reactions) produce electromagnetic radiation across a very broad spectrum – from gamma rays to radio waves, and everything in between. The visible light it sends to Earth across the 150 million kilometres of space [27] in about eight minutes keeps virtually everything on the planet alive from day to day. Why? Because the clever

plants who exist down at the hardworking end of the food chain turn sunlight into something edible for the rest of us [28].

It's remarkable that, after travelling 150 million kilometres in eight minutes, the sun's rays don't feel like taking a nap under a tree. In fact they're ready to energise anything and everything that's in their path – and just one of those things might be a parabolic dish.

Do we all know what a parabola is? If not, take a coin out of your pocket (a rock will suffice for the poor or stingy) and toss it gently from one hand to another.

Congrats: you just made a parabola [29].

A parabola is a two-dimensional shape that stars in maths textbooks and – as you've just discovered – occasionally cameos in the real world as well. Translate a parabola into a third dimension and you end up with a parabolic dish (which probably exists in really difficult maths textbooks, but more usefully in your car headlights, your telescope, satellite dish or at the bottom of your can of soft drink).

Different parabolic dishes do different things, but their claim to fame is that they can refocus a light source and bend it to your will.

Just as a normal dish is a great device for collecting nuts and berries and such, a parabolic dish is a terrific thing to have if you want to collect electromagnetic radiation.

Archimedes's Death Ray is another example of a parabolic dish, or something trying to approximate a parabolic dish. Probably not surprising, given the Greeks' love for a bit of geometry mixed with a bit of wonton destruction.

Ordinary mirrors are dandy reflectors of visual light (a neat trick they accomplish by being very smooth), but of course visual light is just a part of the electromagnetic spectrum. With a few different materials wrought into the shape of a parabolic dish you can be merrily reflecting electromagnetic radiation up and down the spectrum.

Example? The satellite dish on your roof is made from a loose metal mesh but is nevertheless accomplished at the gathering of signals from the satellite and reflecting them to a single concentrated point, whence they are turned into your favourite science television program. In fact, while we're on the subject of reflecting light – let us tell you that you yourself are reflecting some light – it's how we can see you! [31]

[30] This is the SECRET footnote, so shhh, quiet down and listen up. We have some answers (BTW, if you don't know what we're talking about, just forget about this footnote and go on about your reading like nothing happened – or there'll be trouble). First: the spherical Van der Graaff thing. It seems that the round shape makes it hard for a spark to jump off. Pointy things on the other hand make it easy. So in a Van der Graaff generator, a whole bunch of pointy things collect the electrons from the belt, transferring them to the big metal round thing, which can build up a bunch of electrons without letting a spark jump off before you're good and ready. There's a snag though – we also found recent research that suggested that rounded lightning rods were much better at being hit by lightning than sharp rods. Why that should be is anyone's guess. On the matter of pointed permanent magnets – the creation of magnets is a tricky business and best left to the experts. Pointy bits do seem to focus the field, but then MRIs use big doughnut-shaped magnets and they're particularly loopy. What we can say is that if you want to get hold of some truly awesome magnets, then scoot around the web for a site that sells magnets so powerful they come with a warning "You must think ahead when moving these magnets". Think about it. And then YouTube 'neodymium' for kicks.

[31] Okay, we can't see you from where we're sitting, but if you'd invited us to afternoon tea and we'd accepted then just by looking at you our retinas would be bathed in the light that you were reflecting.

Parabolic dishes turn up everywhere, because they're really good at two things. They can turn a dispersed light source (say, the light globe in you car's headlights) into a strong beam of parallel (or at least less dispersed) light that helps you avoid running over kangaroos, sacred cows, moose or frogs on the road at night (depending on where you're living).

But if you're concerned about freezing to death because your aeroplane has crashed in the Andes with a supply of soft drink cans, you'll be more interested in the second ability of the parabolic dish – which is essentially the first reversed.

You can take rays of parallel light from, gosh, well, any stellar object that happens to be handy (yes, the Sun would be a good example, well done everyone on that one) and concentrate those rays into a single point.

Whathewhere?

It's all about the angles of incidence and reflection, baby.

Imagine you've got a parabolic dish and a rubber 'bouncy' ball. Drop the ball on the floor and it will bounce up to your hand and you'll snatch it up and yell 'gotcha!'. However, drop it into the parabolic dish (anywhere but the dead centre) it will bounce out at an angle that will have you searching the living room on your hands and knees.

Now, if you had the time and patience (and supply of 'bouncy' rubber balls) to repeat this experiment a few dozen times, you start to notice that after they bounce in the dish they pass through the same point in space just above it. If you dropped several at once, in fact, they'd have a good chance of hitting one another at that one point.

That point is called the focus, or focal point, of the parabolic dish, and it's really cool because if you do the same experiment with rays of light from the Sun, you'll quickly discover that it's really hot.

Oh yes; because the rays of light from the sun include thermal radiation (some people call it 'heat') as part of their electromagnetic fashion range, and because you're pushing all these rays (and the thermal radiation they contain) into one point, that point is going to get hot. Just how hot depends on the quality of the reflector, but as Tory discovered with the chocolate-rubbed soft drink can parabolic dish, it can get quite hot indeed.

Tory: *"Ow! The thing's burning my hand!"*

See?

Now, the other way to bend light to you will is to use a lens. The particular lens the *Mythbusters* managed to manufacture was made of ice, and it was round.

A round ice lens? Again, it's all about the angles. Just as you burned ants with a magnifying glass when you were a kid (and by the way, the ants haven't forgiven you – they're siding with the frogs), that magnifying glass is essentially made of two slices of glass that might have come from a giant glass ball (but almost certainly weren't).

So, a ball (a 'sphere' to scienticians) of glass would do the same thing. And as long as your ice is nice and clear, it'll do the same light warping thing – and start a fire.

You can make your own water sphere lens with a fishbowl. Take the fish out first, then slap a candle on one side and a white card on the other. You will discover the magic of refraction and magnification all at once! Oh, and the cool 'upside down'-ness of the image on the card (because the rays going through the bowl get bent in opposite directions – you'll see the same if you look into a spoon).

Wacky pop culture reference; when David Blaine (an American street artist … Google him) did that stunt in the spherical glass tank, it was discovered that the bowl was actually focusing enough light and heat to warp the stage it was set on.

[32] That's a cute but frustrating way to answer a simple question, and yet it is annoyingly accurate. After rigorous examination of the elements of her story, the Mythbusters concluded that Lucille Ball can't have either caught or understood radio waves (whether they were bearing simple Morse Code or the fruity jives of Benny Goodman). HOWEVER (and as the episode goes on to relate) there are cases on record of similar things happening. A veteran of the Vietnam War was found to be picking up a radio frequency of 560Hz via bits of shrapnel stuck in his head. The shrapnel was embedded in the bone, and by freakish coincidence could detect and demodulate the signal into sound frequencies that were then transmitted via the bones themselves right into the insides of the ear.

THE MAGIC OF RADIO

When Guglielmo Marconi failed the University of Bologna entrance exam in 1892, few would have guessed that this dud son of a wealthy businessman would soon be responsible for the first communications system utilising invisible waves (or rays) of photons.

Still fewer Bolognians (yes, that's what they're called) would have guessed that 50 years later those same waves of photons (not exactly the same ones) would be broadcasting light jazz and secret Japanese Morse Code straight into Lucille Ball's mouth.

So it's fair (and very satisfying) to say that 'precisely zero' was the number of circa 1892 Bolognians who could have guessed that in the early years of the 21st century a 'television' show, itself broadcast via photons moving at a not dissimilar wavelength to the photons making up the radio waves the Bolognians also hadn't heard of, would attempt to prove that radio waves can be received by fillings in a person's teeth in such a way that their brain can hear the tune they're broadcasting.

Can a brain translate electromagnetic radio waves into music? The answer is, of course, no … but also, and rather more surprisingly, yes [32].

The head and the brain, of course, are outside the purview of this volume, but radio waves are most certainly not. So what does it take to make radio – and even moreso – to catch and understand what it's trying to tell you? Well, it can be surprisingly simple.

Take this exchange for example;

Jamie: *"A pencil and a razor blade and some coil … I mean it seems unlikely that we could actually hear a radio signal from that …"*
Adam: *"Well, right now we're hearing a buzz."*

Okay, let us hear that you're impressed! Yeah, baby!

Because what Adam and Jamie are talking about is a crude version of the classic crystal set – a 'foxhole radio' as they were christened by the poor fellas confined to them (the foxholes you twit, not the radios – you couldn't fit a full-sized soldier inside a radio!).

If you're old enough to know who Marilyn Manson is, then you've probably never heard of either a crystal set OR a foxhole radio, so we might have to wind back the dial even more.

What is radio?

Radio (and television) waves are made up of the more casually-dressed photons who hang out at the 'no worries, she'll be right' end of the electromagnetic spectrum. These waves range in size (properly – 'wavelength') from about one millimetre to 100,000 kilometres.

100,000 km sounds big (alright, alright, it IS big) and this is why the radio telescopes you might encounter in the middle of various nowheres in Australia, India and New Mexico are such oversized bits of kit; 'cos they have to catch 'em. They're a big net for a big fish.

Thankfully for the righteous boffins that get excited about VLAs and Giant Meterwaves, the earth's atmosphere is pretty casual about letting these wavelengths through, although thankfully it makes strenuous efforts to 'bounce' the nastier wavelengths who move around up at the 'Mummy, why is my hair is falling out in handfuls?' end of the spectrum [33].

Lots of things give off radio waves (besides your local station). There's lightning for a start (even static sparks), and all kinds of crazy stars (including ours), interstellar gas clouds, and weird things we don't even know anything about but file in the 'we don't know what's out there' folder.

It's in the hope that some of these 'we don't know' things (which might turn out to be space fleas, Daleks, or vast, multi-dimensional alien dance parties) give off radio waves; that these electromagnetic wavelengths are a useful thing to study in order

[33] Next time you hear some bearded tree-hugger raving about how important the ozone layer is to the breeding environment of green-breasted cocklethrush, tell him from us that the Earth's atmosphere does a lot more than that; it bounces a whole bunch of nasty photons (like those that make up UV-C) that would otherwise zip through your brain and turn you into a pile of radioactive dust faster than the male green-breasted cocklethrush can collect bracken for a nest (of course, the green-breasted cocklethrush would be toast as well).

[34] The story of how this freak fact about static was sussed is a doozie – plug in the names 'Penzias' and 'Woodrow' into your favourite search engine and (as long as it's Google) you'll hit the Nobel Prize website.

[35] This is also why your AM reception crackles up a bunch when there's lightning around.

to hopefully score an invite (to visit the multi-dimensional dance-parties rather than the Daleks).

Okay, hauling it away from Dr Who and back to radio.

A simple radio transmitter can be made from pretty much any old bits of rubbish you have floating around – as long as they include a 9V battery and a coin. Grab an AM radio, switch it on (doesn't matter what frequency) and pop it close to your 9VCOIN radio experiment. Now start tapping away at the two terminals of the 9V with the coin.

…cccrrrzzzppffssstttt… cccrrrzzzppffssstttt… cccrrrzzzppffssstttt… cccrrrzzzppffssstttt…

You ARE listening to the IN sound of 9VCOIN radio!

Don't scoff at static – it's actually quite cool when you consider that the static you see on an unused TV frequency is actually some of the best evidence we have that there was once a Big Bang. Stare at the static and you're staring into the wonder that was the creation of the Universe billions of years ago…neato! [34]

These little momentary chunks of 'radio' blasted out a few centimetres from 9VCOIN are electromagnetic waves that you're sending through the air as a by-product of making electrons flow from one terminal of the 9V, through the coin, and into the other terminal. [35]

This is just the kind of thing that the AM radio is hoping to pick up, because when the electromagnetic wave broadcast from the coin hits the antenna, it tempts a few electrons to move around inside it (as well as in any other bits of metal about the place).

However, the AM radio (unlike just any old bit of metal about the place) has a nifty bit of electronics called a diode, which allows current to flow in one direction but not the other (for this reason it's also called a semiconductor).

When the electromagnetic waves hit the antenna, the signal made by the jiggling electrons is clipped in half, sent to an amplifier (we'll deal with that later) and hence to the speakers (boyo boy, we'll REALLY deal with them later) where you get the magical sound of …

…cccrrrzzzppffssstttt… cccrrrzzzppffssstttt… cccrrrzzzppffssstttt… cccrrrzzzppffssstttt…

Okay, it's not Maria Callas – but what do you expect from 9VCOIN? Their playlist is limited, but it's the same principle that brings you everything from P Diddy to the final overs of Sri Lanka vs. Pakistan at Wankhede Stadium.

If you want to transmit anything other than cccrrrzzzppffssstttt you'll need to put a bit more into the waves your sending out, and fortunately radio waves can be messed around with in a few ways.

The classics are the AM waves that have their amplitudes (the size of the wave) modulated, and the FM waves that have their frequencies (the speed of the wave) modulated (the observant among you will have spotted a correlation here – we're not going to say any more than that). 9VCOIN broadcasts is pulse modulated (PM; essentially on or off) which is good for things like Morse Code, but not much else these days.

Back in the days when Marconi was messing about with radio,

[36] Hertz is a measure of frequency long before it was a company renting cars. One hertz pretty much means 'one per second', so in fact, if the Hertz auto rental company were renting out rental cars at one per second, then Hertz would be renting cars at 1 Hz - which would be pretty funny (if you like your gags really, really lame). Your GPS receiving frequencies at 1500 megahertz means that the waves are oscillating one and a half billion times a second.

he could blurt out whatever taps and crackles he fancied on a range of frequencies simultaneously (we call it broad spectrum – like the sunscreen, and, yes, for the same reason). However, with so many bits of wireless technology lining up for space, they all have to keep to a specific frequency. These are the call signs or station numbers of your AM and FM radio stations, but the crazy thing is that everything from free-to-air TV stations (54 to 88 MHz) to garage door openers (40Mhz) baby monitors (49Mhz) and GPS systems (1227 to 1575 Mhz) have their own little slice of the spectrum [36].

But we're getting ahead of ourselves (how naughty are we?) because we're not planning to start a radio station here; as Adam and Jamie found out in this episode, government regulators make it kinda hard to start doing that just for kicks. All we want is to understand a little more about crazy Lucille and her radio teeth.

The best way to dig on just how simple it CAN be to pick up radio signals (even if we agree that Lucille was doolally, which is what the evidence seems to indicate), is to get back to that foxhole radio. One of the basic and freaky principles of its operation is that it requires NO EXTERNAL POWER. That's right – no battery, no mains, no NOTHING.

Well, not 'no NOTHING' exactly; because it's powered by the energy inherent in the radio waves themselves.

Oh yes – there's power in them there radio waves, and if you've a long enough aerial, there's enough power to hear the dulcet sounds of Nazi propaganda OR armed forces radio, depending where on the razor blade you place your pencil. We know, it's freaky – go with us.

In the 'Free Energy' myth, one of the slightly less than 'Uloida! Uloida! My mother duck is a stoat' crazy theories that the Mythbusters tested was one that relies on a very long antenna that picked up all the 'free energy' being broadcast hither and thither by radio, TV, garage door openers et cetera. And it worked – only problem was it picked up enough power to ALMOST run a very, very small digital watch … (ah yes, but for free!)

Given there's so many signals out there, and the relative simplicity of radio reception, perhaps we shouldn't be surprised at the claims of one sadly departed comedy diva, but instead surprised that we aren't all picking up radio, TV, even each other's mobile phone calls in our various body piercings (perhaps it'll be the wave of the future).

In fact, it can be hard to get away from all the waves we generate. For this reason Adam and Jamie had to drag out a nifty bit of kit called a Faraday Cage [37] for the tests in this episode. A Faraday Cage isn't as kinky as it sounds (not this one anyway). It's a cage where all the sides are made of a conducting material (brass in this case) designed to block all electromagnetic signals.

If you think that such a device is a pretty specialist bit of kit that you'd have no need for in a million years, then take another look at the big box you zap your microwave popcorn in. Oh yes – that's what keeps the microwaves IN. Read on for more…

[37] This is probably going to tell you all about Michael Faraday and his amazing work that linked magnetism to light, but you probably already know all about Faraday don't you? Fer sure dude – he rocked! (he's another chemist, by the way, and he really was a cool dude who told the British Government to sit on a candle when they asked him to make chemical weapons for the Crimean War).

THOSE KILLER MICROWAVES

It seems so ordinary, sitting there on the kitchen counter … waiting until your back is turned … when it becomes … THE KILLER MICROWAVE OVEN.

It's dosh of course – microwave ovens are safe as houses. It's not that microwave ovens can't hurt you; don't ever climb inside one, or drop one on your head. But they're NOT 'nuclear' and they're NOT found in tanning beds.

However, they ARE, basically, radio transmitters.

Eh?

So why is Jamie so excited by what is really a dysfunctional radio station? Because … he's Jamie.

Jamie: *"Microwaves are really cool, you know … I'm gonna, you know, make this super powerful microwave gun that I can heat things up around in my shop with."*

Adam: *"I don't know if he's planned on shielding anything but I don't necessarily want to be in the building when he does it."*

Gawd luv 'im.

The stuff you cook inside a microwave oven is almost as excited as Jamie, but for a very different reason.

Is it because cooking with microwaves was the first step forward in cooking since we started frying food in oil, like, 10,000 years ago? No, but you're getting warm (he he he, 'getting warm' – we crack us up).

Try this fact on for size – microwaves cook food with electromagnetic waves that are about 10 or 11 centimetres in wavelength (yes, in the band referred to as 'radio waves'). So then why do they cook stuff? Your radio doesn't cook stuff.

Interesting point, but, for a start, your radio is receiving the waves, not generating them. If you stood right next to a high-powered radio transmitter it wouldn't do you any favours. Remember, electromagnetic waves are as energetic as aerobics instructors if not more (check over the radio chapter for the bit on crystal sets to reassure yourself, OR just step out into the sunlight and warm yourself up in their warm, energetic glow).

Inside the microwave oven, as the microwaves zoom through the popcorn, they shake up some of the molecules. The molecules that get the best shakeup are the ones that have all their negative charge at one end, and all their positive charge at the other end. To impress your science-y buddies, refer to them as polar, or better yet 'di'-polar molecules (like the poles of the earth, get it?).

As the microwaves zoom past these dipolar molecules, they wiggle around like children's television personalities (if only we could microwave them) as they try to align their negative and positive ends with the waves of electromagnetic energy that's rushing past. But we eat hot, delicious, cooked popcorn, not wussy wiggled popcorn – so what's going on?

What's going on is that the molecules get wiggled pretty rapidly; about two and half billion times a second. That sort of high frequency wiggle builds up a chunk of heat, which transfers to other molecules that are nearby, which transfer it to the bowl itself, and the hands of the sucker who reaches in to picking it up – eventually leading to the

[38] As a prize, look up 'Microwave grapes' in YouTube, but don't do it at home.

3) EXPLODING WATER

kind of 'Yeowwwch' that is heard wherever incautious people and microwaves are brought together.

You can check this out by tossing a cube of ice into a microwave; ice doesn't let molecules gad about like skivvy-wearing lack-brains, so it is much less prone to the heating effect of microwaves than liquid water. It is surprisingly resistant to their charms.

The myth about boiling water in a microwave has raised a little controversy; Adam and Jamie stated categorically in the episode that purified water would not 'boil' in the microwave because it didn't have any impurities to make it boil. Turns out there are some who believe that things called 'microbubbles' have something to do with it, and if you shake up pure water and then boil it you'll be enjoying non-exploded pure-water tea (or other beverage) in no time.

How to decide who's right and who's wrong? Well, hands up if you support Jamie and Adam because they're cool and you're reading a book with their heads on the cover? Okay, now hands up if you support the faceless, perfectionist 'gotta have everything thing just so' anorak-laundering scientificans?

Adam and Jamie win. [38]

Metal things in a microwave are a different story entirely. The spectacular displays of crunched (or Jamie's yet unpatented 'concertina-ed') aluminium foil are almost as spectacular as the kinds of warnings we should probably include that would instruct you not to do such things at home.

Again – why should alfoil (and CDs) do such cool, sparkly things? It's not those weirdo polar molecules again is it? Not this time. This is good old zaps at work.

When you toss some metal into a microwave it's like you're throwing in an antenna – there's a bunch of energetic electromagnetic waves coursing through a hunk of highly conductive metal and suddenly free electrons are dancing about like nobody's business. All that activity winds up as heat, and sometime it does even more than that.

If that bit of conductor you slam in the microwave to make your contribution to the YouTube 'Microwavecrazy' genre happens to receive enough energy to get enough free electrons zotting around inside them, and if there are a few pointy bits [39] on the conductor, and if – you know, just by chance the pointy bits on this conductor happened to be, gosh, close-ish to one another (like in a fork) you could get that energy to arc.

Lo and behold, cue the wild buzzing, the brilliant dance of light … think God about to touch the muscly naked guy in that Vatican ceiling painting by that Italian sculptor (we're sure he meant to add the little arc of electric charge between the two of them) … it's the whole box and dice.

Don't even think about doing this at home. We couldn't abide it if anything went wrong.

What could go wrong?

Well, the fork might not do it for a start (which happened on the show) but the aluminium foil and the CD were priceless! Wicked! Unreal! Outtasite! … groovy!

Don't even think about doing this at home.

Again – why? Well, you could badly damage the thing inside the microwave that's making all those microwaves in the first place.

And what exactly is that?

It's the heart and/or soul of the microwave, the chunky bits of kit that Jamie tried, unsuccessfully (despite a loud buzzing) to turn into some kind of evil super microwave gun.

They're called magnetrons, and we'll find out all about them now, because it is also the heart of radar, which just happens to be the next chapter.

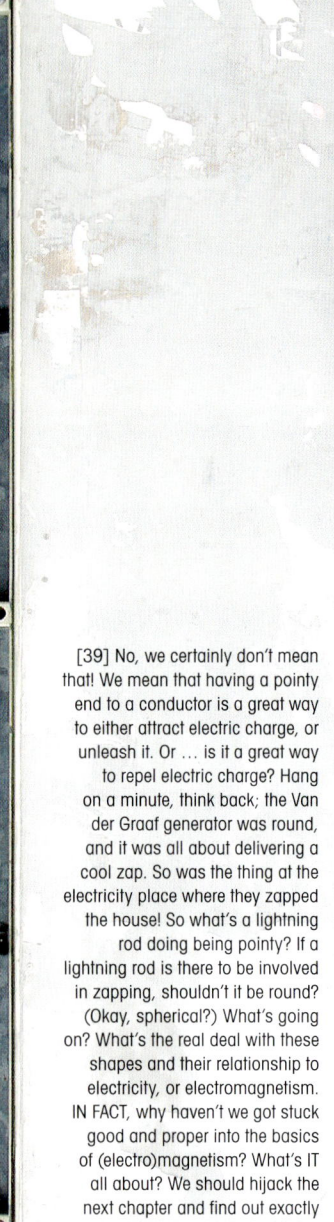

[39] No, we certainly don't mean that! We mean that having a pointy end to a conductor is a great way to either attract electric charge, or unleash it. Or … is it a great way to repel electric charge? Hang on a minute, think back; the Van der Graaf generator was round, and it was all about delivering a cool zap. So was the thing at the electricity place where they zapped the house! So what's a lightning rod doing being pointy? If a lightning rod is there to be involved in zapping, shouldn't it be round? (Okay, spherical?) What's going on? What's the real deal with these shapes and their relationship to electricity, or electromagnetism. IN FACT, why haven't we got stuck good and proper into the basics of (electro)magnetism? What's IT all about? We should hijack the next chapter and find out exactly that. Are you with us? All right! Now, don't tell anyone – they'll not suspect a thing…

WE INTERRUPT THIS PUBLICATION TO BRING YOU AN INTERRUPTION, INSPIRED BY, DEDICATED TO, AND ENTITLED –

THOSE SLAMMIN' MAGNETS

"That's a magnet all right." – Adam

"Gauss is the measurement for magnetic strength!" – Jamie

"…Oh, okay …" – Adam

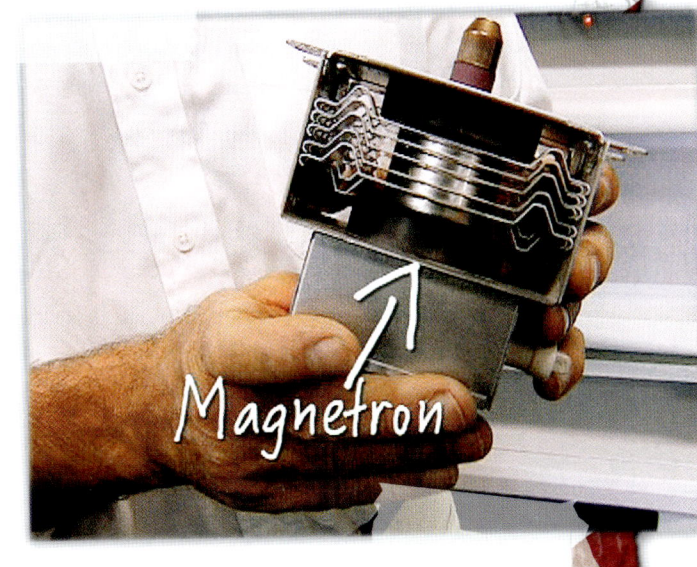

Adam's just being shy – because there isn't a reasonable person alive who wouldn't agree that magnets are some of the slammin'st things on the face of the earth. The fact that kids dig them to the max is proof positive especial, amigo. Show us a happy child, and we'll show you a child with a box full of magnets and a patch of bare earth to hunt for 'ferromagnetic' treasure.

Why are magnet's so damn cool? Is it because a magnet can make the impossible possible? Sure, but it's not just about sticking to the side of the fridge or picking up paperclips a dozen at a time or hovering trains above their tracks at nearly 600kmph (awesome though these tricks are).

Is it just that we LOVED magnets when we were kids and used to sleep with them instead of a teddy bear?

Okay, we've shared too much.

But we're not ashamed – we know we're not alone; magnets are a favourite of the Mythbusters. When faced with the Impossible Heists Myth, Jamie produced some magnets from his own personal collection that were so powerful, they had Adam wincing.

Jamie: "These are not your normal magnet … It'll pull somewhere between 220 and 450 kilograms."

Adam: "So, if you got two of these together, that's like dropping a 450 kilogram weight on your hand."

Jamie: "Yeah. If I put another one here, you would lose your hand."

Magnets are everywhere, inside everything, making it all work. Believe us when we tell you that magnets brush your teeth, buy your lunch, drive you to the hospital and look inside your body because you collapsed in a weird way just after you ate. Your TV, stereo, computer and microwave would all be dead, dead, dead without a cheeky magnet in them.

Actually, electricity, your brain, this planet, the Sun that keeps it alive, all demonstrate the wonder that is magnetism. [40]

Think we're kidding? Think again.

'Electromagnetism', for thus it is called, is one of the big four fundamental forces from which all other forces in the Universe are derived [41]. Some big names have grappled with its mysteries; you may have heard of one, Albert Einstein, the world's most famous patent clerk.

Yet even despite the silver-haired thinkings of old Albert, magnetism is so fundamental to the very nature of everything that we (and by 'we', we mean 'they') still don't even understand it completely (although we – and here, by 'we' we mean 'we' – enjoy it when scientists admit they don't know 100 per cent how things work yet. It's suspenseful.).

Fundamental, infamous and mysterious electromagnetism may be, but there are some things that are comprehensible to you and me; certainly more comprehensible than –

$$1s^2 2s^2 2p^6 3s^2 3p^6 3d^6 4s^2$$

---

[40] Then there's the crazy MagLev trains – check them out on YouTube. Nearly six hundred of kilometres an hour suspended on magnets – hmmmm, magnets…

[41] The other three forces are gravity ('tick' yes, heard of that) and the 'strong' and 'weak' nuclear forces. The strong nuclear force holds the atomic nucleus together, the weak just makes certain things decay in a radioactive way. There you go, that's the forces of the Universe for you – thought there'd be more to it?

[42] Gosh, gee whiz, 'permanent' is such a tough term. It really implies, you know, permanence, and there's really nothing that's so reliable (except death and taxes). Even the toughest 'permanent' magnets out there lose about one per cent of their strength every 10 years. So what's the future like for the really big magnets like the Earth and the Sun? … Are you sure you want to know? Then refer you to the aphorism we used at the start of this rather bleak footnote. Just don't take it out on magnets.

– which is the curious 'electron configuration' of iron, one of just a few elements (along with nickel, cobalt and some wacky members of the 'rare earth element' family, called so despite the fact that they're not so rare (although they are found on Earth (which is hardly surprising, considering where we mostly look for stuff))) that can be made into the 'permanent' magnets [42] that are stuck on your fridge.

Something that we hope to get away with stating point blank is that when electrons move, magnetic fields are generated. This is certainly useful information to hang on to, despite the fact that 'point blank' turns out to be something of a gulf on the sub atomic level, because as the wearers of industrial-strength anoraks out there will know, if you could travel at the same speed as an electron (which is the speed of light, so you can't) you would notice the absence of a magnetic field around the cheeky little electron racing alongside you.

But let's leave that to one side.

For the sake of everything that's not at a speed-of-light, sub-atomic level, when electrons move they create magnetism. So any electric current is creating the same kinds of field lines that you probably remember from science class when you slapped a magnet under a piece of card scattered with iron filings (we loved that one).

Fields, field lines – what's going on here, eh?

The shapes formed by those filings describe a magnet's magnetic field. Magnets have two points where their force is strongest – called the North and South poles. You know that of course; you've probably heard, then, that the earth has them as well.

Actually, the earth's magnetic poles are a bit slippery. They can move around, and they do. They can also swap. And some believe that they can vanish entirely (NASA says it'll never happen), something it shares with the magnetic strip on the back of a credit card.

When Jamie and Adam tested the myth that an eel skin wallet could wipe a credit card (an absurd yuppie myth if ever we EVER heard one) they busted it good and proper [43]. We learned that credit cards are essentially a slice of plastic with a slip of what is essentially VHS tape stuck to them. But how does a little strip of plastic tape get magnetised? It's not made of iron!

Ah, we must pull you up there.

Whether it's a credit card, VHS tape or audio tape (remember those?) it's all about iron, baby. Tiny particles of the stuff affixed to the tape are given a pattern – representing data of some kind – to remember. And they remember that pattern until their iron particles all fall off, or they get spooked by another magnetic field (don't ever leave a VHS tape next to a speaker).

Mythbusters also snuggled up to magnetism when they whaled on the Exploding Tattoo. Tattooing itself we'll get stuck into (hmmm… pun…) in another book, but the chunky apparatus at the centre of the myth is based on magnetism. It even lauds the fact in its name – Magnetic Resonance Imaging.

[43] In fact, it's in our top 10 most busted myths, based on the fact that A) eel skin wallets are mostly NOT made of eel, but of a revolting snotty animal called a hag fish, B) credit cards can stand up to more than 1500 times the magnetic strength of the planet itself, and most of all C) the eel (if it was an eel) is DEAD by the time it's turned into a wallet, and the cool trick it can do with zaps is long since gone.

[44] Confused? Go watch the episode again; Exploding Tattoo – was in the same episode as Quicksand and Appliances in the Bath. Volume 7 as released on DVD. Go and buy it. You get to see Scottie in a hospital robe and Adam in a pith helmet.

How do THEY work? Gosh – do you have a week? In a nutshell then, MRIs use a really, really powerful magnetic field (about 100,000 times as strong as the earth's field) to make the protons inside the hydrogen atoms in the subject (you) line up with the magnetic field like they were the needles of a bunch of compasses.

Then, the MRI swings out its sucker punch – a brief burst of radio waves. These waves flip the protons around, and – here's the good 'image-y' part – when they flip back they send out a teeny burst of radio waves of their own. This radio signal represents the particular kind of tissue they are, and is captured as an image for your doctor to look at and tell you to get more exercise and eat more salad (good advice).

Does an MRI's sub-atomic slapping hurt? No, but you won't be allowed in the room with your glasses, watch, keys, penknife and .. you get the idea. No, unless you've recently swallowed a nail. The MRI magnet is so strong.

BUT there is a snag to magnetic force, as demonstrated *The Exploding Tattoo* as well as *Impossible Heists*. Even the iron-rich inks the build team made up did little more than leave a red welt on a piece of pork [44]. And Jamie's big supermagnetic boots could barely get to grips with the thin metal of an air conditioning duct.

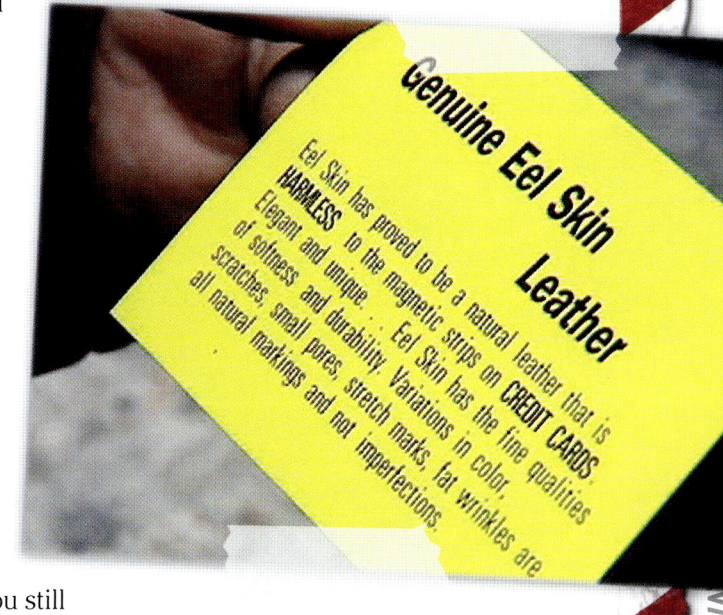

See, the strength of a magnet's pull depends not only on the material, but the mass of that material. Your chunky neodymium 'rare-earth' magnet which you still can't pull off the sledgehammer would not rip a paperclip through your hand – there's just not enough of the paperclip for the magnet to pull on.

How are these crazy magnets made? Simple, they're made by magnets. Okay, so how are those crazy magnet-making magnets made?

The crazy magnet-making magnets are also made by magnets. And yes before you ask …

You make a magnet by first acquiring the kinds of material that becomes magnetic (the magnet-loving community call the stuff ferromagnetic) and then expose it to a strong magnetic field.

Seems suspect? Like to know more? Well gosh junior, if you can just hold onto your pants a little more tightly, we are just about to reveal the secrets of the magnetron.

THE JOY OF RADAR [45]

Before Adam and Jamie began their search for Jimmy Hoffa on – or rather, under – the mysterious 'bump' on the ten-yard line at Giant's Stadium in New York, they did something strangely, spookily sensible.

They believe, as do we, that before messing with anything to do with organised crime it is vital to make a disclaimer; and we stand firmly and squarely with Adam (rather, we stand firmly and squarely behind Adam, whilst wearing anti-tank cologne and bomb-proof undies) when he says:

Adam: *"All we're interested in is taking some science to the myth, whether or not he's here. Don't even want to know who put him here … If someone told me who put him here, I'd forget about it."*

Disclaimer done – on with the science.

How can you search for that is something buried deep underground, and encased in concrete? Before you reply, let us specify WITHOUT digging up the said ground, and demolishing said concrete. You can do that, but you'll end up saying something like…

Adam: *"… I see a pig, oh man, a pig-shaped armhole!"*

---

[45] Are they KIDDING? We were talking about magnets! How did we get pulled back into radar?! I know I know, the magnetron, but what about the lightning rods? The Van de Graaff spheres? Alright, here's the plan; we'll do the research ourselves, right here in the footnote department. Then, at some stage you'll notice a missing footnote number, okay? We can go backwards through the footnotes, so it may have happened already in page-time, so find that missing footnote and we'll be there with the answer to our questions about pointed lightning rods, spherical Van der whatsits, and different shaped magnets and … whatever else we think is required. Shhh – see you there.

[46] Come on, you're not asking for an explanation of the curvature of the Earth? What century were you born in? Look, the Earth is round. However, if you want to know how far away the horizon is (and that's a much more sensible question) let's begin by assuming you're sitting on a beach (nothing but the best in our science fantasies). Now, you'll need two things: a knowledge of ancient Greek geometry, and the distance to the centre of the Earth. So, you should know that when faced with a right-angled triangle (it's corners being your eyes on the beach, the point of the horizon you're staring at, and the molten centre of the Earth itself) you can work out the distance of the side that goes from your eyes to the horizon by adding the squares of the other two sides and taking the square root of the result. Knowing, as you so, that it is 6,378,100 metres to the centre of the Earth (okay, we're assuming all distances from the surface to the centre of the Earth – even at sea level – are equal, which they're most certainly not), and your eye is another two metres above that (no, you're not that tall sitting down, but neither are you right on the shoreline, are you? You're a bit further up the beach!) then … look, this will take forever and suck up valuable page real estate. Let's just say that if you're sitting on the beach two metres above the sea level the horizon is about 3.5 kilometres away, but further if you're higher up, and completely different if you're on another planet.

Nasty.

Instead, let's broaden the question and ask something like this; how do you search for stuff that you can't see? Stuff that might be hidden by night, or by distance, or by fog, clouds or even the curvature of the Earth [46]?

Well, essentially the point is you can only see what you can see, and until about 60 years ago everything else was guesswork, or you sent

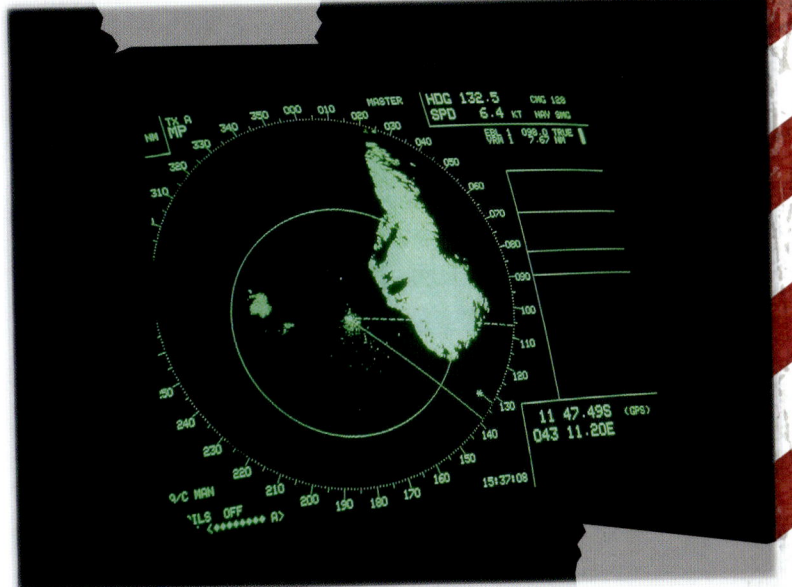

reckless people on dangerous 'scouting' parties, or you used spies, or – of course – maps. But guesses can be wrong, scouts and spies can be killed, captured or bought off, and maps are only useful if you're concerned about things that don't move around much.

And it's the things that move around that are often the things you need worry about – such as a squadron of Messerschmitt BF 109s.

But if it IS a squadron of Messerschmitt BF 109s, then you want to see them when they're still far enough away that you seeing them can be turned into some useful kind of action.

And that's the joy of radar – seeing what you can't see [47].

Radar stands for Radio Detection and Ranging (which is a cheat as far as acronyms go, but a good name nonetheless – better than RaDeRa) and that's a pretty accurate summary of what it does. For a start, it's all about radio waves.

Now, we know that light waves help us see (by lighting things up with wavelengths to which our eyes are sensitive) and radio waves can carry sound, and microwaves (another kind of radio wave) can cook. But how can radio waves see?

It's one of those basic principles of electromagnetic radiation – radio waves, and other electromagnetic radiation, have this habit of reflecting off stuff and generally bouncing around. Visible light reflects off all kinds of surfaces, which is how you can get from your house to your place of business without walking into trees, manholes and semi-trailers.

Whereas humans have eyes calibrated to sense the reflections of light waves, humans had to build machines that would 'see' the reflections of radio waves.

Radio waves reflect when they hit something significantly more dense than the medium they're travelling through (such as a rainstorm floating in the air of the lower atmosphere, in the example of weather radar), and they reflect extra especially well off electrically conductive materials such as the all-metal monocoque of a Messerschmitt BF 109 [48].

However, in the case of the ground-penetrating radar used to spot Jimmy Hoffa shaped cavities in Giants Stadium, the medium is dense and the target is a void. However, fortunately it's the variation that matters, and the variation that is identified when the signal is reflected.

In bedrock, concrete or dry, sandy soils, ground-penetrating radar can 'see' 15 metres underground – that's like the height of a five-storey building. However, in some clay soils (or even in wet conditions) where the electrical conductivity is greater, you might be restricted to just a few centimetres, in which case it's time to reach for the shovel.

To accomplish the task of detection, a radar system must transmit radio waves, and also be able to receive any radio waves that get reflected. Radar systems do this by alternating between the job of transmitting and the job of receiving.

Of course, the bits of radio wave that are reflected back to the radar system are never as strong as the ones sent out. There's also a whole bunch of other fiddly details involved in making the

[47] 'Ultrasound' is another way we've hooked into the visualising potentials of sound, and if you've not had the exhilarating pleasure of seeing your child in the womb thanks to the wonder of ultrasound, then maybe you should consider searching 'ultrasound' on YouTube.

[48] Schluß mit den Messerschmidt-Anspielungen - der Krieg ist seit sechzig Jahren vorbei! Die Alliierten waren schließlich auch keine Waisenknaben müssen wir Euch an Dresden erinnern? (Danke Wilfried)

[49] It's been suggested that some of you might like to know more about how 'stealth' planes, boats etc. are designed, and why they look so 'clunky-yet-funky'. Imagine you're a pulse of radio waves, and you're bent on hitting a target aircraft and reflecting enough of yourself back towards the radar station. Well, your best bet is to hit an aircraft that's made up of lots of rounded shapes (like a jumbo jet), because there's a better than even bet that some part of that shape will be pointing back at the radar station. However, if it's all flat angles, or worse, angles that are all internal (like the inside corner of a cube), then you'll just bounce off into the Earth or into space, anywhere but the radar installation. Make sense?

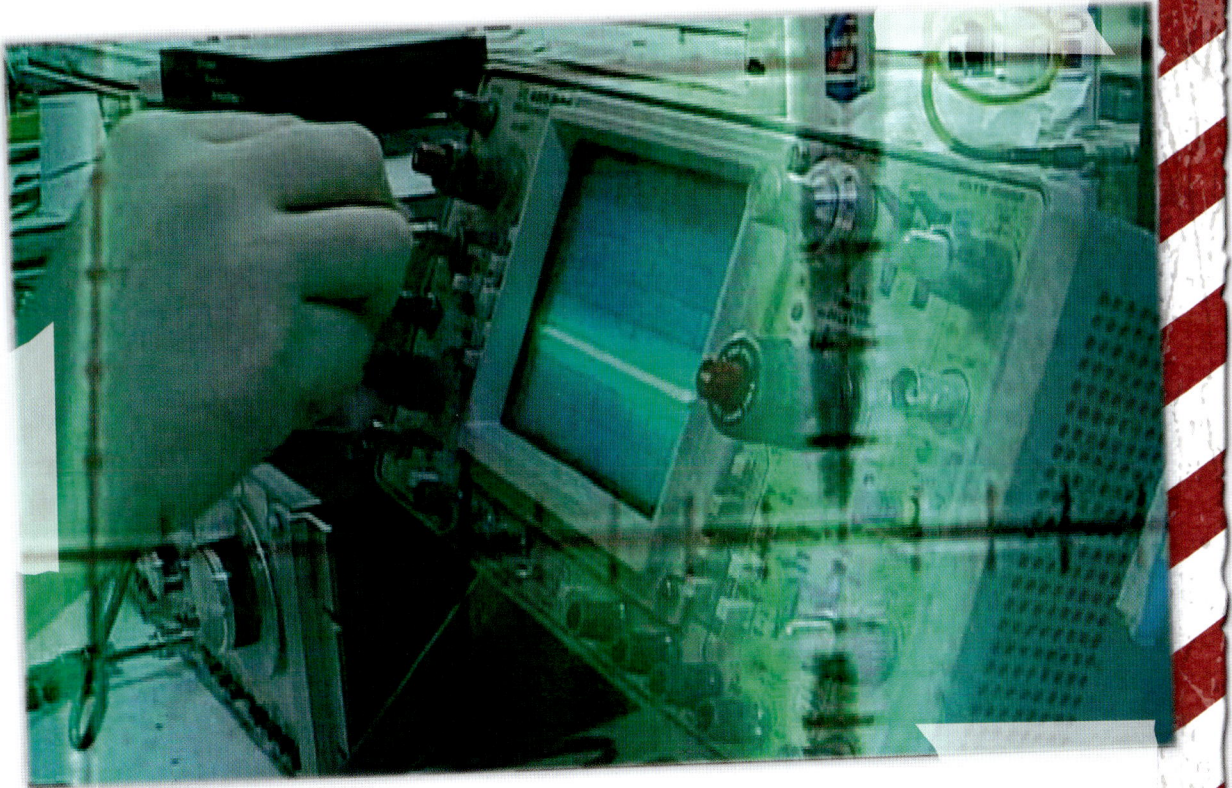

reflected signal into useful information: like how the object to be detected reflects the radio signal (different shapes can disguise an object's 'radar cross-section'; look at those weirdo 'stealth' planes for instance [49]), and how the radar spots all sorts of things that your anti-aircraft guns aren't interested in hitting with their three-inch armour-piercing rounds (like a flock of birds, your own planes, the rain), and how the presence in the area of other conflicting electromagnetic noise that could be accidental or a deliberate jamming attempt, and much, much more.

Now, back in the early days of radar, those fiddly details all meant a lot of work for people with thick glasses, slide rules and grease pencils. Thankfully computers are enslaved to that drudge work these days, yet all of the fiddle that went into making the success that was and is radar (rather than the duff series of screw-ups that might otherwise have become known as ra-'duh') could well be the reason people still think today that jamming radar should be easy. Since law-enforcement bodies have stated using radar and a closely related tracking system called lidar (see breakout) to enforce speeding regulations, the good citizens of the world have a strong motivation to retain the belief that jamming radar is easy.

'Lidar' is much the same as radar but uses light or even lasers, and can be built into much smaller and much cooler 'gun' units, allowing law-enforcement officials to do that "I'm gonna bust ya" strut when they approach your car, having pulled you over.

Yet, however imaginative and multifarious the systems of avoiding police radar (and lidar), none of the methods investigated by Tory, Scottie and Kari were worth either the trouble of making them, or the immense daggyness of driving the public highway with them attached to your vehicle.

Don't know what we're talking about? Watch the episode.

These imaginative and multifarious methods fall into three camps; deflecting the radar / lidar waves away from your car, fixing something to your car that will not reflect the waves, and using other kinds of electromagnetic radiation to jam the waves.

Now, it's important to point out that all these things CAN BE DONE by – for example – military research institutions with a few billion dollars in their back pocket. It's a little harder to pull such systems together in your carport the afternoon before an illegal drag race [50]. Hence the abject failure of all the systems put to the test that warm spring day on a highway in California.

[50] Of course, there's a bunch of stuff you can just buy that can either detect and / or jam police radar and lidar. None of them are 100 per cent foolproof because – despite what you read to the opposite in the popular press – the police and the people who organise the police are not fools. Lidar, for example, uses pulses of light that are at their most effective when aimed at your licence plate rather than your dash, where your detector / jammer is sitting. Also, police have at their disposal radar detector detectors, which have probably resulted in the invention by some anorak of a radar detector detector detector, and so on, ad infinitum. Probably the most salient point is that technologies that detect or jam police radar / lidar on behalf of drivers who like to use the public highways to cannon along at life-threatening speeds are illegal in most sensible places around the world. And so they should be.

We did promise you the inside skinny on the magnetron – the heart of both radar and the microwave oven. It's important, we think, to say that magnetrons are not the sorts of things you or I really need to know much about.

In fact, ask a non-anorak wearing person in the street 'what is a magnetron?' and they'll probably tell you to go back to your comic books. But let us oblige your quest and tell you that a magnetron uses a strong permanent magnet to influence the electrons exuded by a cathode to cycle past tubular cavities and induce a high frequency radio field which … blah blah blah … ends up cooking your food or spotting your enemies.

Hoorah.

Let's move on.

The Basics of Waves

Chapter 3

There's waves and there's waves.

There's the kind of waves that make you grab your board and run for the water like a hungry penguin, the kind of waves that emanate from your throat when you hit that water at 12°C, and then there's the kind of waves that push through hundreds of kilometres of solid rock to knock down buildings and make, in turn, really BIG waves.

And they're all different, yes?

Well, no. These waves are all essentially the same.

We'll let Adam explain.

 *"… all sound is a set of oscillating pressure waves." – Adam*

Did that make all the sense in the world? Should he say it again?

 *"… all sound is a set of oscillating pressure waves." – Adam*

Okay, okay, alright, yes, thank-you, there is more to it than that. My, you're all budding scientists now aren't you? We're mmm-gosh, OH so proud.

Ahem.

Sound waves, water waves and earthquakes are called examples of mechanical waves [51]. What we were talking about in previous chapters were electromagnetic waves (or rays – but let's not start that again). Yes – the perceptive among you have noted that they are spelled differently, and the reason is they're very different things.

The key difference is this; mechanical waves can only travel through a medium.

No, we're not talking about the kind of medium who's usually a woman with only one name who works in a darkened room that smells like damp tobacco and who vibrates alarmingly when shown $20 (although the reason she's called a medium might finally become obvious once you read the next sentence).

[51] – The word 'mechanics' comes from the Greek for 'machine', or perhaps 'pertaining to the workings of a machine'. In fact, the title 'mechanics' was slapped on the science we now call physics back in the time of ye olde famous scientist Sir Isaac Newton, because it was well known to even the local axle-fitter that the Universe was nowt but a giant machine. But be warned, if you were to stand up in your next physics class and proclaim that the great Newton was a mechanic, be prepared for some even-bigger anorak to say 'In fact, Gordon... ' (for this is your name in our little fantasy classroom) …'In fact, Gordon, Sir Isaac was a mechanician.' At which point the only thing Gordon can do to regain his / her (your) pride is flush the poor lad's head down the nearest lavatory.

[52] – In fact Shakespeare refers to a certain rough 'salt of the Earth' type as a 'mechanical' in a certain play about a certain dream. Don't tell us we never give you any kulcha.

[53] – The movie poster for 'Alien' was entirely correct when it blared 'In space no one can hear you scream', but probably soon fell into inaccuracy when huge space explosions were made to seem audible from some distance away – another inaccuracy being there's no oxygen in space to combust in an explosion.

A medium is a material that allows the transmission of waves from one place to another.

Now, these mediums ('media' to get all grammatical on your ass) can be anything from solid rock to the wanton summer air (ahhh, Shakespeare [52]). BUT you don't get mechanical waves in a vacuum (like outer space), for they cannot propagate ('do stuff') if there's nothing for them to push around [53].

It's like this; if you dangle a piece of rope and give one end of it a wiggle, that wiggle travels down the length of rope until it runs out of rope, by which time your cat is probably dangling from the end of it. The point is, without the medium – the rope – your rope wave cannot continue to explore the Universe like an electromagnetic wave could.

And thus it is for sound. Sound needs to move through something to eventually reach an ear that will listen to and appreciate it (whether that sound is Ella Fitzgerald singing 'Old MacDonald', or delighted giggles from the two-year-old in a dancing frenzy who's listening).

Mostly, that medium is air, but sound also travels very nicely through other things as well. In fact, sound travels through water even faster and with greater efficiency (ask a whale).

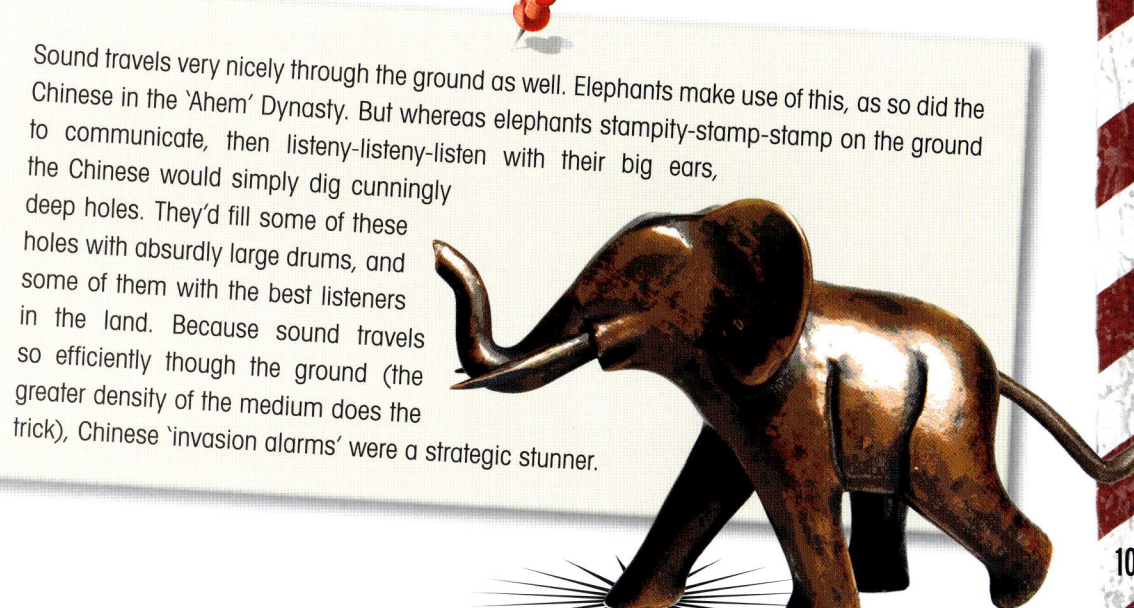

Sound travels very nicely through the ground as well. Elephants make use of this, as so did the Chinese in the 'Ahem' Dynasty. But whereas elephants stampity-stamp-stamp on the ground to communicate, then listeny-listeny-listen with their big ears, the Chinese would simply dig cunningly deep holes. They'd fill some of these holes with absurdly large drums, and some of them with the best listeners in the land. Because sound travels so efficiently though the ground (the greater density of the medium does the trick), Chinese 'invasion alarms' were a strategic stunner.

Now, if instead of a rope you had a guitar string attached to, well, a guitar, you could pluck that string and lo, make a sound. But what, exactly, have you done?

The guitar string is vibrating very rapidly; if you touch it you'll feel them briefly before those vibrations dissipate into your finger [54]. But if you don't touch it, then each vibration moves a little bit of air, and this changes the pressure in the air as it does so. If this happens enough times in a second, you get a sound.

[54] - Feels nice doesn't it? Don't you wish you'd kept up those music lessons? You could be rocking out right now for a bunch of groupies, instead you're tucked up with a good book.

[55] – If you liked that then you'll love the fact that the shockwave associated with the eruption (another mechanical wave) actually circled the Earth several times.

The sound lasts for as long as the energy you've put into your plucked string is able to move bits of air around before dissipating (both from the string and from the air). It might be 10 metres in the case of a delicately plucked guitar string, or many thousands of kilometres in the case of the eruption of Krakatoa at 10:02AM on August 27, 1883, which could be distinctly heard 3100km away in Perth, Australia, and 4800km away on the island of Rodrigues near Mauritius…

How's that for a fact? [55]

How we hear the sound is a whole other story, suffice to say it involves evolving tiny things in your ears to catch vibrations of a certain spectrum, then turn them into electrical impulses and injecting them into your brain.

What's this certain spectrum? Just as your eye can only put certain wavelengths of light to work, your ears can only decipher certain wavelengths of sound. Anything below about 20Hz (cycles a second, ja?) or above 20,000Hz is lost to our ears – but that's just us. Bats, elephants and dolphins do VERY clever things with sound that's way, way out of our league.

Because sound is really waves of pressure travelling through the air, it can actually have a considerable impact beyond the sweet song, hilarious joke or the fingernails on a blackboard that they deliver into your ear.

Beyond a certain loudness (call it amplitude, if you will) a sound can blow your head off. Could it demolish a car?

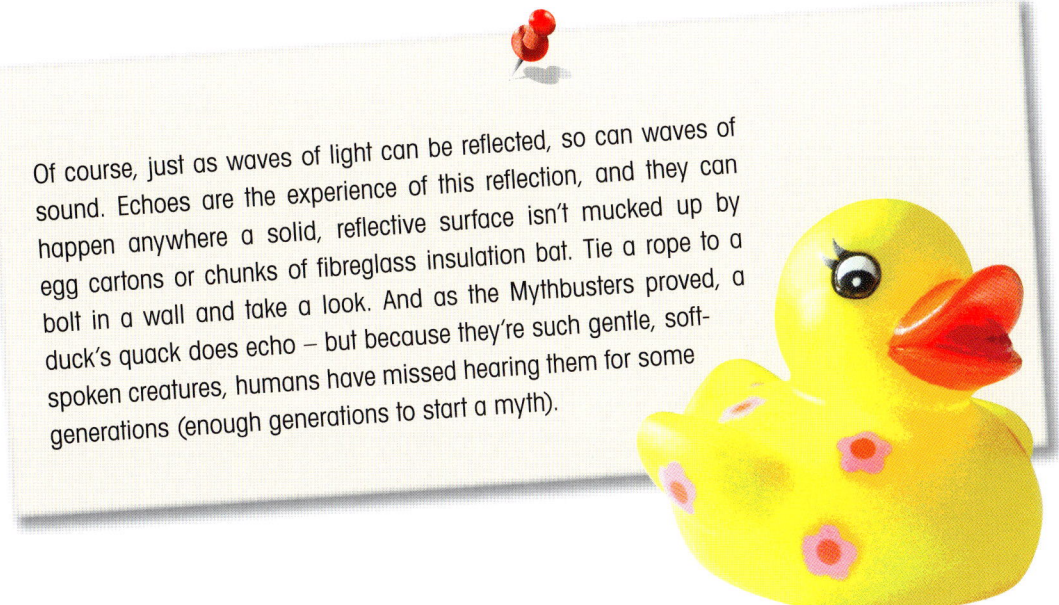

Of course, just as waves of light can be reflected, so can waves of sound. Echoes are the experience of this reflection, and they can happen anywhere a solid, reflective surface isn't mucked up by egg cartons or chunks of fibreglass insulation bat. Tie a rope to a bolt in a wall and take a look. And as the Mythbusters proved, a duck's quack does echo – but because they're such gentle, soft-spoken creatures, humans have missed hearing them for some generations (enough generations to start a myth).

THE WEFT OF THE WOOFER

"So do we bring it up to a certain speed and just run it till it busts or do we try and like ... ?" – Jamie

 *" ... I wanna crank it and just send it to the moon."* – Adam

Nice work Adam – rarely has the entire Mythbuster ethos been so elegantly described.

Ahh, speakers. Try to list five areas of human endeavour more completely and utterly choking with guff than that of the loudspeaker. [56]

When Alexander Graham "I'll Call You Later" Bell patented the very first loudspeaker as part of his 'Do you think it'll ever catch on?' telephone in 1876 [57], who would have put much money on the availability of the VAST array of sound projection equipment available today, variously marketed with funky shapes, new materials, endless statistics and phrases like 'point sources', 'multi-drivers' and 'bending wave transducers'?

Hmmm ... bending wave sounds good ... how much for a pair?

Okay, okay, maybe you would have put money on it (and so would we). And why not? In their ability to mimic any and every sound you could imagine (within reason [58]), speakers are entirely and utterly remarkable in almost every way.

And it's all about pushing air.

You can make a speaker out of almost anything – in fact, we're fairly sure that the kinds of minds that saved the astronauts on Apollo 13 by building a CO_2 scrubber from old copies of the Astronaut's Union Monthly could make a decent quality speaker out of a child's toilet, 'twisty' ties, and a couple of decent fridge magnets (not the bendy ones).

[56] - How about muscle cars, brain surgery, trainspotting, quantum string theory, economics...this is just too easy. We could add plumbing and winemaking as well.

[57] – A quirk of fate could easily have led to one Elisha Gray inventing the telephone. 'How's that?' you ask, in your quest for water cooler factoids. This is one of those times when we're going to say 'Wiki him!' and leave the rest up to your reading skills.

[58] – Outside that 'reason' are, for example, sounds so loud that to make them would destroy the equipment, and sounds way above our hearing threshold. Most speakers give in at around 20,000Hz.

The key to a speaker is being able to precisely control how you're pushing and pressurising the air to replicate sounds, be they John Zorn, Kamahl or the Drummers of Bora Bora (your choice). And, once again, it's magnets to the rescue (at least in traditional speakers – we'll leave the whole piezoelectric thing to another book – or even another series).

A speaker uses a permanent magnet and an electromagnet (in a speaker it's called the 'voice coil') to create a perfect frictionless environment. Inside here the real work is done, as the voice coil, which is attached to the speaker cone, can vibrate happily in and out of a groove in the permanent magnet to push around the kind of air that will recreate sounds with as much perfection as a plastic potty can.

Far, far beyond the plastic potty end of the spectrum, anyone who's gone shopping for speakers knows that there is endless variety out there – and as far as the hi-fi anoraks are concerned, if you're not spending a year's executive salary you ain't getting anything worth having.

But that's not what Adam and Jamie were interested in when they took one look at a large German car and thought 'I wonder if we could destroy that with sound?'

Their speaker – all 51 inches of it – was directly driven by a chunky diesel engine, and is probably the world's first and, so far, only diesel-powered speaker. Did the world need it? Does the world need 1,324,397 flavours of ice-cream? Would people buy diesel-powered speaker-flavoured ice-cream?

The answers of course are yes, yes, and yes again.

But how can a diesel engine drive a speaker? If you're following, you'll now know that a speaker is just pushing air – and it doesn't need to be hooked into a rare recording of 19th Century Hungarian soprano Katharina Klafsky to do that. All it needs to do is go up and down.

This it managed to do with enough efficiency to bust the car's sunroof with a sound at 16Hz (sixteen cycles per second) and at 161decibels. Down at 16Hz you can't hear the sound, but you can <u>feel</u> its pressure (certainly at 161 decibels you could, as the experiment demonstrated).

[59] – The guy was dead when the 'bel' was named (early 1920s) – so how could he push for that kind of honour? And while we're on the subject of units of measurement, have you ever stopped to consider what unit of measurement might best be named after YOU? It's a spooky thought, and probably one only ever considered by excited 19-year-old physics PhDs when they get a whiff of a breakthrough.

So then decibels, everyone's favourite scale of sound (known as 'dB' to their closest friends), that's a simple 'walk in the park' kind of thing, yes? You bet! As long as you find logarithmic ratios simple.

You do, do you? Well, explain it to the rest of us.

" ……. "

We didn't think so.

The decibel is best known for its work measuring sound, but it also crops up in electronics, optics, earthquakes, wherever a cunning logarithmic device is needed.

Yes – we said earthquakes.

Let's go back a step or two; a decibel is really just a bel in disguise. And the 'bel' is a corruption of 'Bell' – Alexander Graham "Call Me" Bell (or rather, the vast corporation he spawned [59]).

So the (deci)bel is one tenth of a bel; interesting, but not getting us any closer to what it is, or why it's hanging around with those logarithms you've read so much about in the Standard Poodle-Owning Longboard Rider's Monthly International Gazette of Clips, Trims, Tides and Swells.

Now, hold on to the book tight, 'cos we're about to jump into a bit of maths.

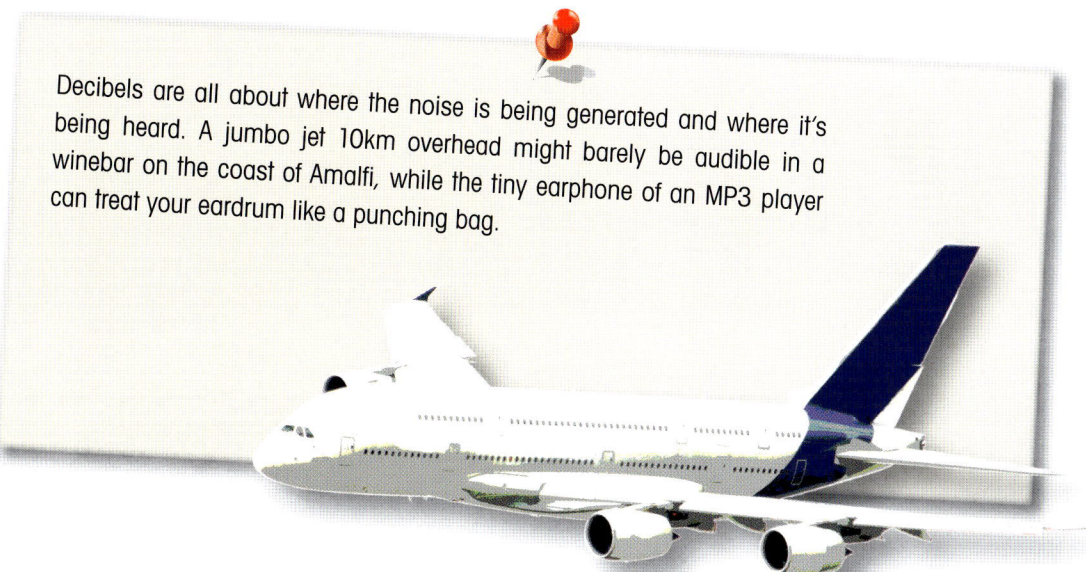

Decibels are all about where the noise is being generated and where it's being heard. A jumbo jet 10km overhead might barely be audible in a winebar on the coast of Amalfi, while the tiny earphone of an MP3 player can treat your eardrum like a punching bag.

Logarithms are a way of expressing numbers that can otherwise get very, very big and unwieldy and use up a lot of zeros. You might be aware of the classic scientific notation that would express two billion as 2×10^9. Neat, smart – the kind of numbering system that you could introduce to your own daughter.

The reason that logarithms (logs to their friends) are used in the expression of loudiosity is that it seems that the human ear can hear across a vast range of sounds from very, very quiet things – like a dastardly bloodsucking mosquito – to terrifyingly loud sound made by the industrial-strength power tool with which you despatch the mosquito because you hate the damn things so much you wish they were all dead, and damn the environment.

Okay, we're sharing too much again, but the thing is this: if you expressed these sound pressures as a ratio (the ear-splitting right-on-the-threshold-of-permanent-damage sound to the barely-audible-but-unmistakably-a-mozzie-out-for-blood sound) it would be more than a million to one.

[60] – If you think we're going to draw a picture of a piano keyboard and point out middle C you've got another thing coming. And that 'another thing' is that we won't point out C6 or C8 either, OR explain what C6 and C8 mean. Go and ask a musician – or better yet BUY a piano, LEARN how to play it, TAKE a degree in applied acoustics and finally, truly, utterly dig the world of sound.

And because the raw power in a sound wave is proportional to the square of the pressure that it exerts on the air, THAT ratio is more than a trillion to one. Don't put the book down. We apologise for the maths.

It goes down like this; if you double the amount of power you're putting into making a sound, you get nowhere near double the volume. This is why the boffins with slide-rules developed a unit of measurement that took account for these big changes in loudness with which we fill our lives – trying to keep it all under control.

The upshot is that it takes a WHOLE BUNCH (yes, we've stepped away from the maths at the moment) more energy to drive up a sound wave from 121 to 122dB than from 3 to 4dB; know what we mean? Each step is one step beyond the last.

Alarmed? Confused? It's disconcerting to be confronted with the complexities of sound, probably because sound is such an everyday experience. You're making a bunch of sounds right now (and so are we) but you don't have to do any kind of calculations to do so.

Content yourself with the fact that one of very cool things about sound is that when you get up, up, up in the dBs (WAY above the 120dB, where you will pop an eardrum, and WAY over the paltry 161dB achieved by Jamie and Adam's MythWoofer) and you're approaching 200dB, you're not really talking about a sound any more.

You're talking about a very serious and essentially inaudible shockwave – like the shockwave made by a nuclear bomb.

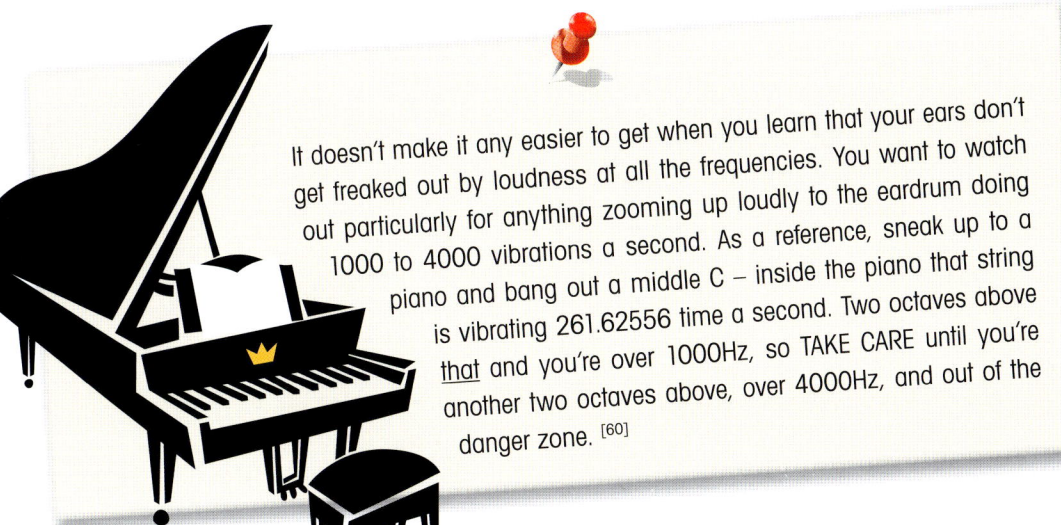

It doesn't make it any easier to get when you learn that your ears don't get freaked out by loudness at all the frequencies. You want to watch out particularly for anything zooming up loudly to the eardrum doing 1000 to 4000 vibrations a second. As a reference, sneak up to a piano and bang out a middle C – inside the piano that string is vibrating 261.62556 time a second. Two octaves above that and you're over 1000Hz, so TAKE CARE until you're another two octaves above, over 4000Hz, and out of the danger zone. [60]

And once you're knocking down buildings with 'sound' you're ready to understand why earthquakes are measured with a system much like decibels – they're waves as well; waves of pressure through miles and miles of rock. That 'Richter Scale' you hear about is really a decibel scale in disguise. Okay – now you're ready for the secret world of mechanical resonance.

IN THE END, IT'S ALL ABOUT THE VIBE

This is our final chapter. Been great y'all, and we like to think we've saved the best for last.

In the course of their busting adventures, Jamie and Adam have tackled not one, not two, but at least [61] FIVE myths that relate directly to one very extraordinarily cool phenomenon.

"And so the theory is that the tone you hear when you hit it is the same tone that will potentially break it?" – Adam

That's exactly right, if a little obscure. Before you do anything else, read this footnote [62] (promise we're not trying to sell you anything).

Do you get it yet? Try another quote.

"The goal is to get the bridge to tune to about three hertz, which is about three beats per second, which is about a soldier's marching cadence. Then once we get that we put our frequency oscillator on it, basically a big weight that hits the bridge at the right moment, bang … bang … bang, and we watch the bridge go wiggy every time we hit it. It should be pretty cool." – Adam

[61] – Correct as at the time of publication – who knows what's happened by the time this book is in your hot little hand? (and why is your hand so hot any way? Actually, never mind…)

[62] – Okay, cool. Google Tacoma Narrows footage. Do it!

[63] – Of course, the episode that dealt with swings – 360 Degree Swing Set – was short on mentions of mechanical resonance, and big on attaching rockets to things – for that is what Mythbusters does best.

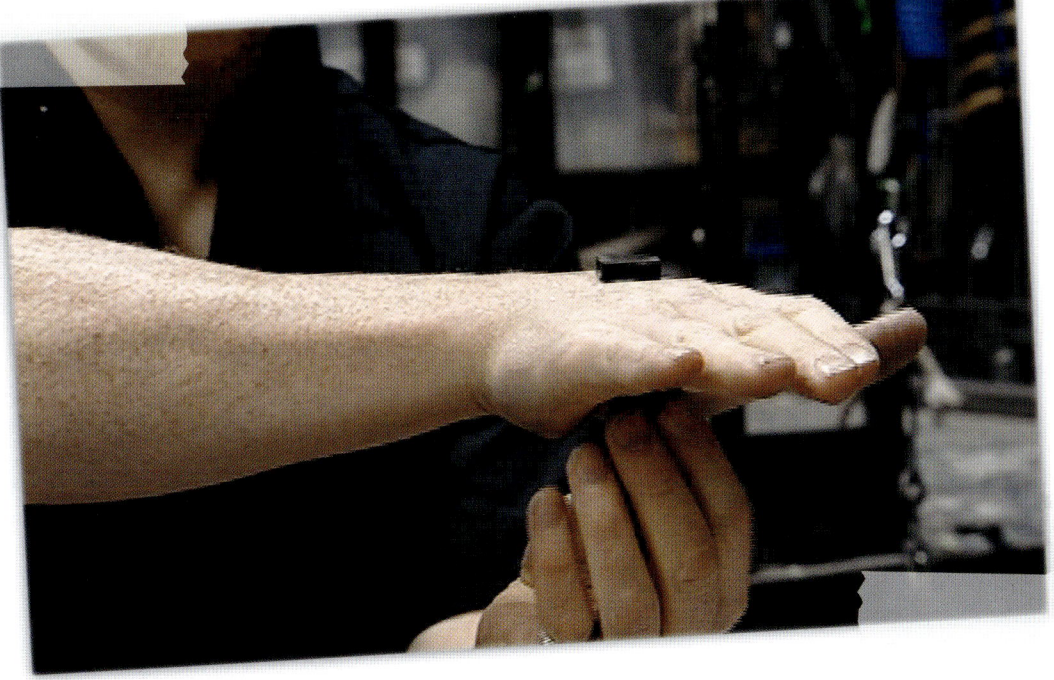

Tuning bridges to make them go wiggy? It's tempting to think that it's Adam who's going wiggy, but the truth is, he's 100 per cent in tune with a truly fantastic phenomenon. In fact, this phenomenon is so 'in tune' with the Mythbuster's ethos it could almost be renamed 'The Mythbusters Principle'.

Yet it is called mechanical resonance, and it is best demonstrated in your local playground.

Recall, if you will, your sun-filled summer days of many years ago (or not quite so many, depending on your age or your sobriety). Bestriding the local park was a childhood classic – the swing set. From about a year old, we all experienced the exhilarating speed and gravitational fluctuation brought about thanks to – yes – mechanical resonance. [63]

How? You know when you're making a swing go higher, you have to time it just right – so as to maximise the swing you get? Kari had it down pat.

Kari: *"Whoa."*
Tory: *"You went over a horizontal."*
Grant: *"Your swing technique is excellent."*

Why is it so? Well, a mechanical system (like the local swing set) absorbs more energy when you are 'in tune' with its 'natural rhythm'.

If that sounds more the kind of hippy talk spouted by your Auntie Rainbow than a scientific principle from a rigorous almanac of facts [64] then think again.

And the fun news is this; a mechanical system is – well – pretty much anything. It can be a bath. Ever pushed the water up and down a bath and noticed that sometimes grand 'bath tsunamis' will form and slosh liberal amounts of dirty bath water over the floor? All good, clean fun, and a prime example of mechanical resonance.

There's a place on the East coast of the USA and Canada called the Bay of Fundy and it boasts the biggest tides in the world – 17 metres! It's due to a curious coincidence that causes <u>tidal</u> resonance. See, the bay is so long that it takes the waves going in at the start of the tide 12.4 hours to go all the way to the end of the bay, and then come all the way back again. That 12.4 hour time span just happens to be the same as the time between one tide and the next. By a quirk of fate the length of the bay is 'in tune' with the 'natural rhythm' of the tides … spooky, spooky, spooky!

But wait – mechanical resonance is not all about messing about in the bath and pushing kids on swings, there's some top quality destruction as well! [65]

Let's take a big thing like a bridge – now, you'll know from walking or driving over them that even a bridge is subject to vibration. But you only strike a problem when those vibrations get 'in tune' with a bridge's 'natural rhythm'.

A bridge with natural rhythm? Now, you may never have seen a bridge at a dance party, but natural rhythm is no respecter of size or materials, and once you hit it – spectacular things happen.

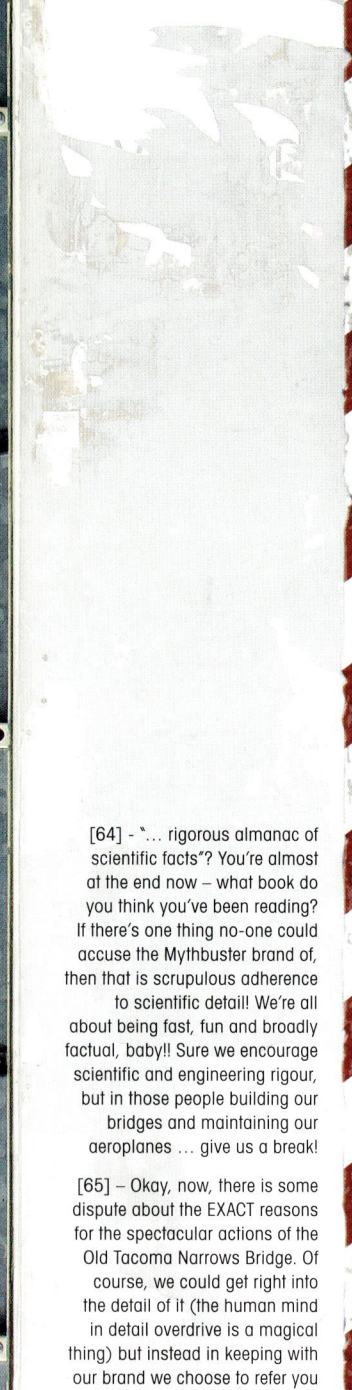

[64] – "… rigorous almanac of scientific facts"? You're almost at the end now – what book do you think you've been reading? If there's one thing no-one could accuse the Mythbuster brand of, then that is scrupulous adherence to scientific detail! We're all about being fast, fun and broadly factual, baby!! Sure we encourage scientific and engineering rigour, but in those people building our bridges and maintaining our aeroplanes … give us a break!

[65] – Okay, now, there is some dispute about the EXACT reasons for the spectacular actions of the Old Tacoma Narrows Bridge. Of course, we could get right into the detail of it (the human mind in detail overdrive is a magical thing) but instead in keeping with our brand we choose to refer you to footnote [64].

Now, just because they failed to happen in the myth of the 'Break Step Bridge' (and not for want of trying) does not mean it's not a worthy myth. In fact, if you ever join the army (it's fun for a while) you'll still probably be ordered to 'break step' when marching over a bridge. Who wants to take the risk of falling 70 metres into a raging torrent of white water when you could be raining down death on the enemy?

Tesla's Earthquake Machine was another story.

Nikola Tesla is one of the most dynamic, distinctive and dramatic figures in the history of science – if not humanity. The list of his inventions and accomplishment stretches from speaking a dozen languages to inventing AC current (and yes, he's got a scale named after him – the 'Tesla' is the unit of magnetic field). So awesome and fascinating was Tesla that some people (freaks, sure, but <u>people</u>) believe he was from the future. He wasn't, but it won't be long before some enterprising screenwriter turns his life into an award-winning movie starring Leonardo DiCaprio as Tesla and Paul Giamatti as Edison (any Hollywood producers out there?).

When Jamie and Adam began to fiddle with Tesla's Earthquake Machine, there were giggles and even snarkyness; how could a simple oscillator (the oscillator was in this case a little machine that thumped a rhythm at a chosen frequency) threaten to bring down a multi-storey building, as Nikola 'Future-boy' Tesla claimed?

But once their homemade 'Earthquake Machine' oscillator was strapped to a fairly substantial bridge, Jamie and Adam were singing another tune.

Jamie: *"Oh my God – it feels like a big semi trailer truck is rolling right by us right now."*
Adam: *"And that's only six pounds of weight moving 25 times a second."*
Jamie: *"It actually makes me a little concerned."*

[66] – Toss 'resonance' into YouTube – there's a very interesting rice show that looks pretty and makes you think (and there's precious little of that on YouTube).

But why, why, why? Why does a bridge, a bath, a swing have some kind of 'natural rhythm'. What is this resonant frequency? How do you find it?

Well, one method is to sing.

Eh?

Crazy as you think it sounds, if you sing at something with a resonant frequency in the audible (or singable) range, you might hear that something sing back at you. Try it by sticking your head inside a piano and singing up and down the scale like Bruno the avant-garde jazz pianist. You know you want to! And you'll be sure of some mechanical resonances.

That's just what the Mythbusters did to stake a world first.

Bringing the boys from Meyer Sound and then rock singer Jaime Vendera into the myth, the Mythbusters set their sights on being the first people ON RECORD to shatter a crystal wine glass with only the power of the human voice.

It was a classic myth – most of us have heard that breaking a glass by singing at it was possible, yet none of us had ever tried it and posted it on YouTube [66].

Singing just the right note at a glass could make it wobble – and <u>seriously</u> wobble. Catch the high speed camera shots; who would have thought a wine glass had that kind of wobble in it?

And if the French military and Japanese police are to be believed, even the human gut can be influenced by resonant frequency. And although the episode dedicated to the Brown Note revealed it to be bustable, even with the Meyer Sound lads in the mix, mechanical resonances (resonant frequencies) are crazy stuff.

Why are they there? What benefit do they serve? Why should stuff simply destruct when struck with air that's under pressure at a certain frequency?

Look, to be honest – and after much searching through bits of guff about the behaviour of particles in quantum field theory, equations that look like they've been developed by a Rhesus monkey with Crone's disease and a bunch of writing that was REALLY weird – we've got to tell you, we can't work it out.

If there is the occasional bit of resonance – whether it's in crystal wine glasses, musical instruments or electrical circuits (oh yes, they've got it too) then we are jolly thankful that there are some good vibes floating around the place.

It's a strange and complicated world out there. Many – if not most – of us only understand the world to the level at which we feel comfortable to 'leave it at that'. Thankfully there <u>are</u> people willing to ask the tough questions, then pour hot wax over the answers to get to the truth. These people are scientists, and the best of them make the world the kind of remarkable place that we love to love.

So – if you want to know the answer to why resonant frequencies exist, or the truth behind lightning, or magnetism, then get out there and look for it yourself, 'cos for the moment we're done here with our 'entertaining and informing' shtick.

But know this – there is much to be learned that is unknown and much to be relearned that was not very well learned in the first place. If you have time in your life, get yourself into some science – it's the only story that really means a damn.

There are people who believe that the whole planet has a resonant frequency. Called the Schumann Resonance, it's – well – jolly complicated, but it suggests that there are links between low frequency electromagnetic radiation, lightning, global climate change and well, gosh, a bit of just about everything. Never heard of the Schumann Resonance? Don't worry, you're not the only one.

We'll Meet Again…

Is it over already? Gosh, we had so much more to say.

We wanted to tell you more about gamma rays, and the frying egg sound of a high voltage current, about stud-finders and why they're not meant to be used in bars, and about a new US Military weapon based on a heat ray, but called an 'Active Denial System' for reasons best known to the Pentagon.

Then there's the things we probably should have got into – like the Doppler effect, that actually stretches light or sound depending on

[67] And goodbye from us here in the footnote department; look out for the little numbers! They're the one's that count.

the relative speeds of the source and the subject. We'd loved to have spent more time on magnets and mechanical resonance (especially as it relates to electronics) and … well, pretty much everything.

But that's it – we're done [67].

DO NOT DESPAIR. There's more to come. In planning are books that delve behind the many myths on the human body, the ancient world, gadgets, cars, explosions, weapons and OH so much more. All you have to do is BUY THEM in sufficient quantities that will keep the ball rolling.

In fact, to bring you all along for the ride we're going to push for the creation of a special website or even just a blog; something that will let all of us talk about all these sciencey things and more with only the interruption of the occasional advertising pop up (people have to be paid somehow you know). Look out for it.

So that's it – we offer no final gratuitous advice, quotes or pat slogans.